Electronic and Electrical Servicing

Consumer and commercial electronics

Second Edition – Level 3

Ian Sinclair

and

John Dunton

AMSTERDAM • BOSTON • HEIDELBERG • LONDON • NEW YORK • OXFORD
PARIS • SAN DIEGO • SAN FRANCISCO • SINGAPORE • SYDNEY • TOKYO

Newnes is an imprint of Elsevier

Newnes is an imprint of Elsevier Ltd
Linacre House, Jordan Hill, Oxford OX2 8DP
30 Corporate Road, Burlington, MA 01803

First published 2002
Reprinted 2003
Second edition 2007

British Library Cataloguing in Publication Data
A catalogue record for this book is available from the British Library

Library of Congress Cataloguing in Publication Data
A catalogue record for this book is available from the Library of Congress

ISBN: 978-0-7506-8732-4

For information on all Newnes publications
visit our web site at www.books.elsevier.com

Typeset by Charon Tec Ltd (A Macmillan Company), Chennai, India
www.charontec.com

Working together to grow
libraries in developing countries

www.elsevier.com | www.bookaid.org | www.sabre.org

ELSEVIER   BOOK AID International   Sabre Foundation

Transferred to Digital Printing in 2007

Contents

Preface to the second edition – Level 3

This new edition of *Electronic and Electrical Servicing* reflects the rapid changes that are taking place within the electronics industry. In particular, we have to recognise that much of the equipment that requires servicing will be of older design and construction; by contrast, some modern equipment may require to be replaced under guarantee rather than be serviced. We also need to bear in mind that servicing some older equipment may be totally uneconomical, because it will cost more than replacement. With all this in mind, this new edition still provides information on older techniques, but also indicates how modern digital systems work and to what extent they can be serviced.

This volume is intended to provide a complete and rigorous course of instruction for the core units of Level 3 of the City & Guilds Progression Award in Electrical and Electronics Servicing – Consumer/Commercial Electronics (C&G 6958). It follows on from the Level 2 book (as do the chapter numbers), which covers all the core units and two of the option units at this level (ISBN 978-0-7506-6988-7).

Acknowledgements

The development of this series of books has been greatly helped by the City & Guilds of London Institute (CGLI), the Electronics Examination Board (EEB) and the Engineering & Marine Training Authority (EMTA). We are also grateful to the many manufacturers of electronics equipment who have provided information on their websites.

Ian Sinclair
John Dunton

Unit 1

Outcomes

1. Demonstrate an understanding of reactance, resonance, transformers and transfer and the practical application of these components and circuits

2. Demonstrate an understanding of semi-conductor devices, displays and transducers and the practical applications of these components.

26 Sine wave driven circuits

Capacitors and inductors in direct current (d.c.) circuits cause transient current effects only when the applied voltage changes. In an alternating current (a.c.) circuit this is a continuous process. Capacitors in such a circuit are therefore continually charging and discharging, and inductors are continually generating a changing back-electromotive force (emf). Circuits containing resistors, capacitors and inductors are described as **complex circuits** and an alternating voltage will exist across each component proportional to the magnitude of current flowing through it so that a form of Ohm's law still applies.

In the explanation that follows, the symbols V' and I' are used to mean peak a.c. values (Figure 26.1) of alternating signals. Root mean square (r.m.s.) values are represented by V_{rms} and I_{rms}; that is, $I_{rms} = I'/\sqrt{2}$. The symbols v and i represent instantaneous values of a.c. signals and V and I will have their usual meaning of d.c. values. So, we may write $v = V' \sin (2\pi ft)$.

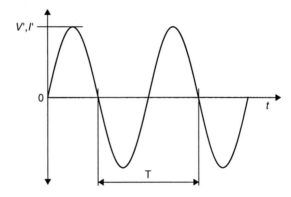

Figure 26.1 The peak value and period of a sine wave of voltage or current

For a resistor in an a.c. circuit, $V' = R \times I'$ and the value of resistance found from the variant form of this equation, $R = \dfrac{V'}{I'}$, is the same as the d.c. value, V/I. In a capacitor or an inductor, the ratio V'/I' is called **reactance**, with the symbol **X**. Because this is a ratio of volts to amperes, the same units Ω (ohms) are used as used to express d.c. resistance, but it must be remembered this is a frequency-dependent reactance and not a resistance.

A capacitor may, for example, have a reactance of only $1\,K\Omega$ at a given frequency, but a d.c. resistance that is unmeasurably high. An inductor may have a d.c. resistance of 10 ohms, but a reactance of $5\,\mathrm{K\Omega}$ or more. The reactance of a capacitor or an inductor is **not** a constant quantity, but depends on the frequency of the applied signal. A capacitor, for example, has a very high reactance to low-frequency signals and a very low reactance to high-frequency signals, as indicated in Figure 26.2.

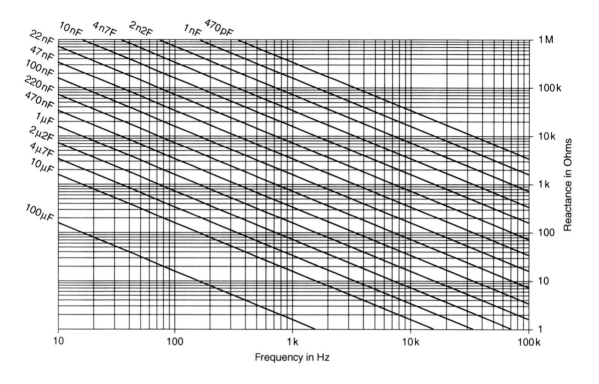

Figure 26.2   Chart of capacitive reactance at audio frequencies

The reactance of a capacitor V'/I' can be calculated from the equation:

$$X_C = \frac{1}{2\pi fC}\ \text{ohms}$$

where f is the frequency of the signal in Hz and C is the capacitance in farads. Figures 26.2 and 26.3 show the values of capacitive reactance for a range of different capacitors calculated for a range of frequencies. These charts are intended as a guide so that you can quickly estimate a reactance value without the need to make the calculations.

The reactance of an inductor varies in the opposite way, being low for low-frequency signals and high for high-frequency signals. Its value can be calculated from the equation:

$$X_L = 2\pi fL\ \text{ohms}$$

where f is the frequency in Hz and L is the inductance in henries. Figures 26.4 and 26.5 show the values of inductive reactance which are found at various frequencies.

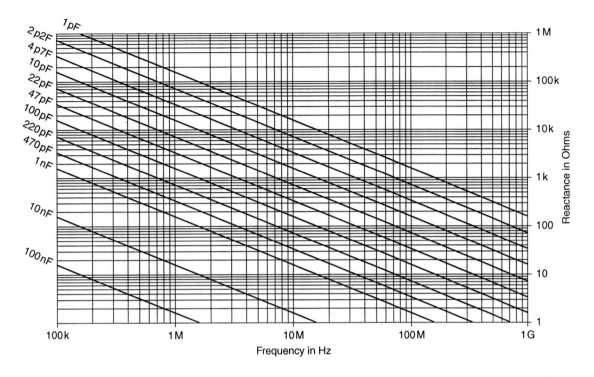

Figure 26.3 Chart of capacitive reactance at radio frequencies

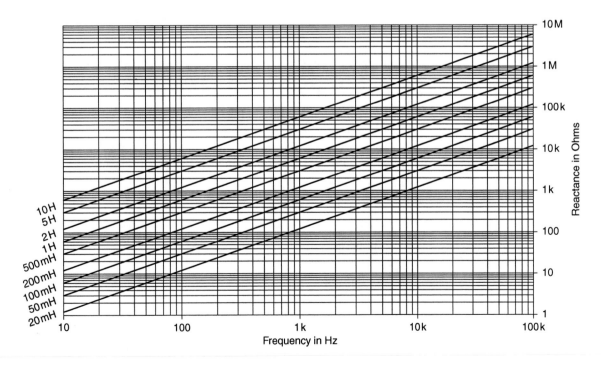

Figure 26.4 Inductive reactance at audio frequencies

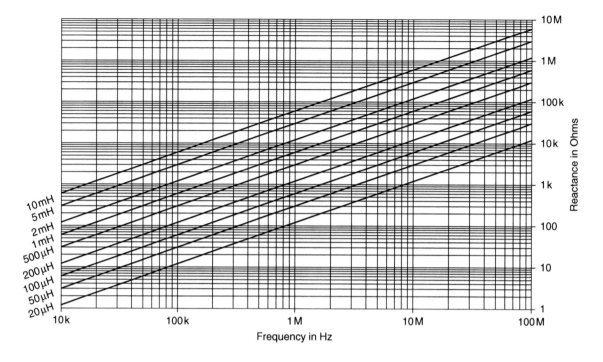

Figure 26.5   Inductive reactance at radio frequencies

Note: inductors are now less common in circuits other than power supply and radio (transmission and reception) applications. The increasing use of integrated circuits (ICs) and digital circuitry has made the use of inductors unnecessary in a very wide range of modern applications.

Practical 26.1

Connect the circuit shown in Figure 26.6. If meters of different ranges have to be used, changes in the values of capacitor and inductor will also be necessary. The signal generator must be capable of supplying enough current to deflect the current meter which is being used.

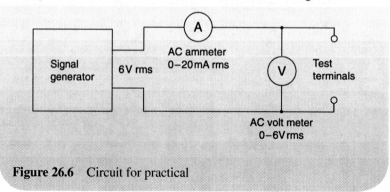

Figure 26.6   Circuit for practical

(Continued)

Practical 26.1 (Continued)

Connect a 4.7 μF capacitor between the terminals, and set the signal generator to a frequency of 100 Hz. Adjust the output so that readings of a.c. voltage and current can be made. Find the value of V'/I'' at 100 Hz.

Repeat the measurements at 500 Hz and at 1000 Hz. Tabulate values of $X_C = V'/I'$ and of frequency f.

Now remove the capacitor and substitute a 0.5 H inductor. Find the reactance at 100 Hz and 1000 Hz as before, and tabulate values of $X_L = V'/I'$ and of frequency f.

Next, either remove the core from the inductor or increase the size of the gap in the core (if this is possible), and repeat the measurements. How has the reactance value been affected by the change?

There is another important difference between a resistance and a capacitive or inductive reactance and this can be demonstrated as follows. With the aid of a double-beam oscilloscope, the a.c. waveform of the current flowing through a resistor and the voltage developed across it can be displayed together (Figure 26.7). This shows that the two waves coincide, with the peak current coinciding with the peak voltage, etc. If this experiment is repeated with a capacitor or an inductor in place of the resistor, you will see from the figure that the waves of current and voltage do not coincide, but are a quarter-cycle (90°) out of step.

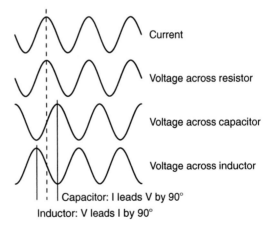

Figure 26.7   Phase shift caused by a reactance

Comparing the positions of the peaks of voltage and of current, you can see that:

- For a capacitor, the current wave leads (or precedes) the voltage wave by a quarter-cycle.
- For an inductor, the voltage wave leads the current wave also by a quarter-cycle.

An alternative way of expressing this is that, for a capacitor, the voltage wave lags (or arrives after) the current wave by a quarter-cycle, and for an inductor, the current wave lags the voltage wave by a quarter-cycle.

The amount by which the waves are out of step is usually defined by the **phase angle**. The current and voltage waves are 90° out of phase in a reactive component such as a capacitor or an inductor.

A useful way to remember the phase relationship between the current and voltage is the word C-I-V-I-L, meaning C–I leads V; V leads I–L. The letters C and L are used to denote capacitance and inductance, respectively.

If we take a few measurements on circuits containing reactive components we can see that the normal circuit laws used for d.c. circuits cannot be applied directly to a.c. circuits.

Consider, for example, a series circuit containing a 10 μF capacitor C, a 2 H inductor L and a 470 ohm resistor R, as in Figure 26.8(a). With 10 V a.c. voltage, V' at 50 Hz applied to the circuit, the a.c. voltages across each component can be measured and added together: $V'_C + V'_L + V'_R$. You will find that these measured voltages do not add up to the voltage V' across the whole circuit.

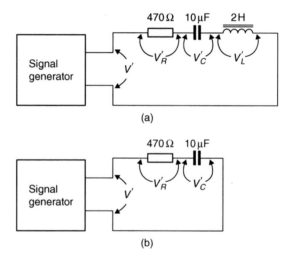

(a)

(b)

Figure 26.8   Series circuits: (a) RLC, and (b) RC circuit for practical example

Practical 26.2

Connect the circuit shown in Figure 26.8(b). Use either a high-resistance a.c. voltmeter or an oscilloscope to measure the voltage V'_R across the resistor and the voltage V'_C across the capacitor. Now measure the total voltage V' and compare it with $V'_R + V'_C$.

The reason why the component voltages in a complex circuit do not add up to the circuit voltage when a.c. flows through it is due to the phase angle between voltage and current in the reactive component(s). At the peak of the current wave, for example, the voltage wave across the resistor will also be at its peak, but the voltage wave across any reactive component will be at its zero value. Measurements of voltage cannot, however, indicate phase angle. They can only give the r.m.s. or peak values for each component, and the fact that these values do not occur at the same time cannot be allowed for by meter measurement. The result is that straight addition of the measured value will inevitably give a wrong result for total voltage, because of the time difference.

Phasor diagrams (often also called **vector diagrams**) are one method of performing the addition so that phase angle is allowed for. In a phasor diagram, the voltage across a resistor in an a.c. series circuit is represented by the length of a horizontal line drawn to scale. Voltages across reactive components are represented by the lengths of vertical lines, also drawn to the same scale. If all the lines are drawn from a single point, as in Figure 26.9(a), the resulting diagram is a phasor diagram that represents both the phase and the magnitude of the voltage wave across each component.

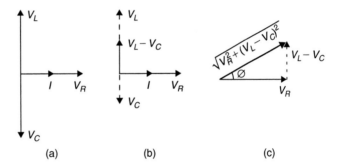

Figure 26.9   Phasor diagrams for complex series circuits: (a) relationship between V and I for a series RLC circuit, (b) combining V_L and V_C, and (c) finding the total voltage across the circuit

To represent the opposite effects that capacitors and inductors have on the phase, the vertical line representing voltage across an inductor is drawn vertically upwards, and the line representing voltage on a capacitor is drawn vertically downwards. By convention, an inductive reactance and the voltage across it are considered to be positive. A capacitive reactance and its voltage are considered to be negative.

The phasor diagram can now be used to find the total voltage across the whole circuit. First, the difference between total upward (inductive) and total downward (capacitive) voltage is found, and a line is drawn to represent the size and direction of this difference. For example, if the inductive voltage is 10 V and the capacitive voltage 7 V, the difference is 3 V drawn to scale in the direction of inductive reactance. If the inductive voltage were

10 V and the capacitive voltage 12 V, the difference would be 2 V drawn to scale downwards in the capacitive direction.

The net reactive voltage so drawn is then combined with the voltage across the resistor in the following way. Starting from the point marking the end of the line representing the voltage across the resistor, draw a vertical line, as in Figure 26.9(b), to represent the net reactive voltage in the correct direction, up or down. Then connect the end of this vertical line to the starting point (Figure 26.9c). The length of this sloping line will give the voltage across the whole circuit, and its angle to the horizontal will give the phase angle between voltage and current in the whole circuit.

The total voltage can be found by Pythagoras' rule, since the phasor relationship is that of a right-angle triangle, so the total voltage is

$$V'_T = \sqrt{V'^2_R + (V'^2_L - V'^2_C)}.$$

A complex circuit which contains both resistance and reactance possesses another characteristic which is of great importance. This characteristic is known as **impedance**, symbol **Z**. Impedance is measured in ohms, and is equal to V'/I' for the whole circuit. Its value varies as the frequency of the signal varies.

When impedance is present, the phase angle between current and voltage is less than 90° in either direction. This phase angle can be found most easily by using the phasor diagram in a slightly different way, to form what is known as the **impedance triangle**.

In a phasor diagram constructed this way (Figure 26.10), separate lines are drawn to represent the resistance R, the reactance X and the impedance Z. In a series circuit, the length of the horizontal line represents the total value of resistance in the circuit, and the vertical line its net value of reactance upwards as before, for predominantly inductive reactance (a), and downwards for predominantly capacitive reactance (b).

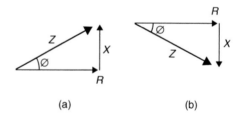

(a)   (b)

Figure 26.10   Impedance triangle: (a) predominantly inductive, and (b) predominantly capacitive

With the values of R and X known and the angle between them a right angle, the Z line can be drawn in, representing the impedance value of the whole circuit. The angle of this line to the horizontal is the phase angle between current and voltage in the circuit.

Another way of working out the relationships between R, X and Z in a complex circuit is to express them by two algebraic formulae:

$$Z = \sqrt{(X_L - X_C)^2 + R^2} \quad \text{and} \quad \tan\varphi = (X_L - X_C)/R$$

where Z = total impedance, X_L = inductive reactance, X_C = capacitive reactance, φ = phase angle, and R = the resistance of the circuit as a whole. A pocket calculator covering a reasonably full range of mathematical functions can now be used to work out the values of circuit impedance and phase angle.

Filters

Filters are circuits that are designed to separate out a range of frequencies of interest that are present in any waveband. Because these are frequency dependent, the circuits must contain at least one reactive component.

In the low-pass filter (LPF) shown in Figure 26.11(a), the inductor has a low series reactance at low frequencies so that these pass easily to the output and are developed as the output signal V_{out}. In Figure 26.11(b), the capacitor has a very high reactance at low frequencies and acts as the load to develop the output signal V_{out}. At high frequencies the low reactance of C effectively short-circuits the output signal.

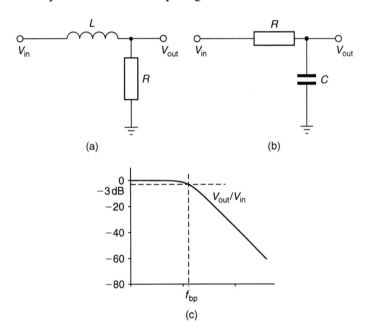

Figure 26.11   Low-pass filters: (a) LR, (b) CR, and (c) typical response

Figure 26.11(c) shows the amplitude of the output signal plotted against frequency for both the LR and CR circuits and this also represents the variation of circuit impedance with frequency. The $-3\,dB$ or half power point of the frequency response represents the **break point** where the circuit reactance is equal to its resistance value. This point is defined by the formula

$$f = 1/(2\pi CR) \text{ Hz} \quad \text{or} \quad f = 1/(2\pi(L/R)) \text{ Hz}$$

and is therefore dependent on the circuit time constant.

For the high-pass filters (HPFs) shown in Figure 26.12(a, b), the resistors and reactors are interchanged to produce the opposite effect. Figure 26.12(c) again shows the attenuation effects and the variation of circuit impedance, with the break point calculated as before.

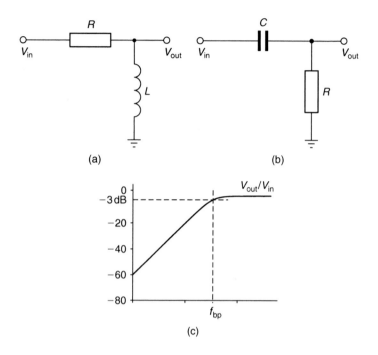

Figure 26.12   High-pass filters: (a) LR, (b) CR, and (c) typical response

For both of the first order LPF and HPF circuits with a single pair of components shown above, the straight part of the attenuation slope falls away at 6 dB per octave (a doubling or halving of frequency).

Practical 26.3

For an LPF and an HPF as shown in Figures 26.11 and 26.12, set up a circuit so that the signal current through the filter can be plotted against frequency. How do these results compare with the shape of the attenuation characteristics?

A circuit that selects a band of frequencies is described as a **band-pass filter (BPF)** and one such circuit can be constructed as shown in Figure 26.13. This is effectively a cascade of an LPF and an HPF and its attenuation characteristic is shown in Figure 26.13(b). The attenuation slope at both ends is still −6 dB per octave. The two −3 dB break points of the individual low- and high-pass sections are calculated from the equations

$$f_{lf} = 1/(2\pi C_1 R_1) \text{ Hz} \quad \text{and} \quad f_{hf} = 1/(2\pi C_2 R_2) \text{ Hz}$$

and if the stages did not load each other the bandwidth of the circuit would simply be the difference between these two frequencies. This is rarely the case for practical values and in effect the high- and low-pass corners are pulled towards each other by the loading effects. This type of filter can suffer from quite large in-band insertion losses, particularly if the RC sections are chosen to minimize load effects; it is often useful to put a buffer amplifier in between the stages to restore the losses and isolate the stages from each other.

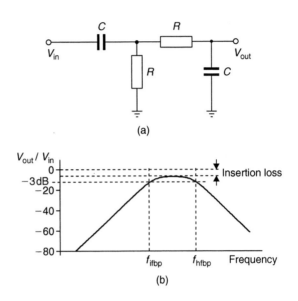

(a)

(b)

Figure 26.13   (a) Band-pass filter circuit, and (b) typical response

A **band-stop filter (BSF)**, which has the opposite characteristic of the BPF, cannot be conveniently constructed using the cascading principle shown above. For these circuits it is more usual to use second order filters as shown in Figure 26.14(a). This particular circuit is referred to as a twin or parallel T device, where the ratios of the component values are chosen as:

$$C_2 = 2C_1 \quad \text{and} \quad R_1 = R_2/2$$

so that the maximum attenuation occurs at $f = 1/(2\pi R_1 C_1)$ Hz.

Resonance

The values of capacitive reactance, X_C, and inductive reactance, X_L, are frequency dependent but have opposite slopes. At low frequencies, X_C is large and X_L small, a situation that becomes reversed at high frequencies. There must therefore be some frequency at which $X_C = X_L$. This frequency is called the **resonant frequency**, or the frequency of **resonance**, of the LCR circuit in question. Its symbol is f_r.

A phasor diagram drawn for a series LCR circuit at its resonant frequency will clearly have a zero vertical component of reactance (Figure 26.15). The

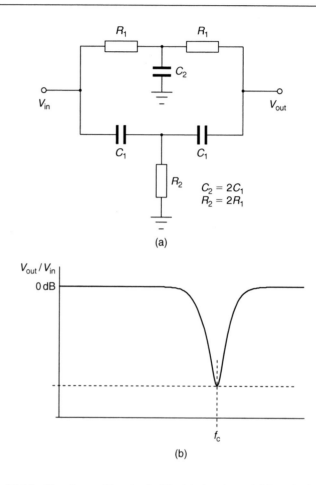

(a)

(b)

Figure 26.14   Band-stop filter (twin-T): (a) circuit, and (b) typical response

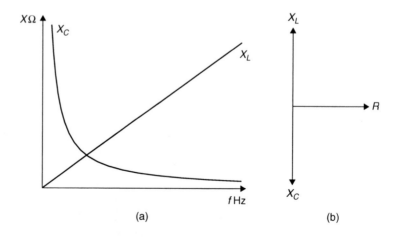

(a)   (b)

Figure 26.15   Resonance in a series LCR circuit

impedance of the circuit will therefore be simply equal to its resistance. The same conclusion can be reached by working out the impedance formula:

$$Z = \sqrt{(X_L - X_C)^2 + R^2}$$

At its resonant frequency, therefore, an LCR circuit behaves as if it contained only resistance, and has zero phase angle between current and voltage.

Practical 26.4

Connect the circuit shown in Figure 26.16 with component values as follows: $R = 1\,K$, $C = 0.1\,\mu F$ and $L = 80\,mH$. The resonant frequency of the circuit is about 1.8 kHz. Set the signal generator to 100 Hz, and connect the oscilloscope so as to measure the voltage across the resistor R. This voltage will be proportional to the amount of current flowing through the circuit, because $V = R \times I$. Now increase the frequency, watching the oscilloscope. The resonant frequency is the frequency at which current flow (and therefore the voltage across R) is a maximum. Note this frequency, and the value of the amplitude of the voltage across R at the resonant frequency. Measure the voltages across L and across C by connecting the oscilloscope across each in turn. Note the value of these voltages. Finally, use the oscilloscope to measure the voltage across the whole circuit. Construct a phasor diagram for the voltages across R, C and L and confirm that this produces an answer for the total voltage. (Remember that the oscilloscope itself will disturb the circuit to some extent, and that the resistance of the inductor has not been taken into account in your calculation.)

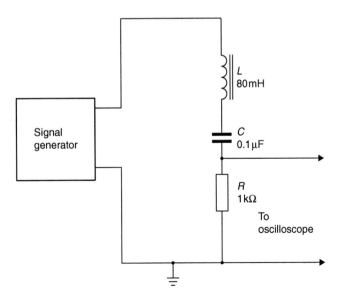

Figure 26.16 Circuit for practical exercise, series resonance

In a series-resonant circuit, the current flow at resonance in the circuit will be large if the voltage across the whole circuit remains constant. Therefore, a large voltage will exist across each of the reactive components. In general, the voltage across both the capacitor and the inductor will be greater than the voltage across the whole circuit at the resonant frequency.

The ratio V'_X/V'_Z, where V'_X is the voltage across a reactor and V'_Z is the voltage across the whole circuit, is called the **circuit magnification factor**, symbol Q, which can be very large at the frequency of resonance.

The frequency of resonance for a series circuit can be calculated by using the formula:

$$f = \frac{1}{2\pi\sqrt{LC}}$$

where L is the inductance (henries), C the capacitance (farads), and f is the frequency (hertz).

Example: What is the resonant frequency of a circuit containing a 200 mH inductor and a 0.05 mF capacitor?

Solution: Substitute the data in the equation, taking care to reduce both L and C to henries and farads, respectively. Use $L = 200 \times 10^{-3} = 0.2\,\text{H}$ and $C = 0.05 \times 10^{-6} = 5 \times 10^{-8}\,\text{F}$. Take 2π as being approximately 6.3. Then:

$$f_r = \frac{1}{(6.3\sqrt{0.2 \times 5 \times 10^{-8}})} = 1587\,\text{Hz}$$

Find (in MHz, to two decimal places) the resonant frequency of the series combination of a 220 pF capacitor and a 15 μH inductor.

A circuit consisting of inductance, capacitance and resistance in parallel will resonate at a frequency given approximately by the same equation that was used for series resonance. However, in a practical circuit, the resistive component is more likely to be the inherent resistance of the inductor; in which case the circuit now consists of C in parallel with L plus a series resistance R that represents this component's losses. The frequency of resonance for this circuit is given by:

$$f = \frac{1}{2\pi}\sqrt{\frac{1}{LC} - \frac{R^2}{L^2}}$$

If R is small, as is usually the case, the R^2/L^2 component can be neglected and the resonance frequency formula reverts to that for series resonance. The **lossy term** R^2/L^2 is related to the Q factor (or quality factor, $Q = 2\pi fL/R$) of the inductor, so that if Q is higher than about 50 the simplified formula is accurate enough for most applications.

At the frequency of resonance, a parallel resonant circuit behaves like a high value of resistance, L/CR, which is called the **dynamic resistance** or impedance. Again, at resonance, the phase angle between voltage and current is zero.

Practical 26.5

Connect the parallel resonant circuit of Figure 26.17 to the signal generator. Find the frequency of resonance, which for the component values shown will be about 2250 Hz, and note that at this resonant frequency, the voltage across the resonant circuit is a maximum. Now connect another 0.047 μF capacitor in parallel with C, and note the new frequency of resonance. Remove the additional capacitor, and plot a graph of the voltage across the resonant circuit against frequency, for a range of frequencies centred about the resonant frequency. Observe the shape of the resulting curve, which is called the resonance or response curve. Now add a 10 K resistor in parallel with the resonant circuit, and plot another resonance curve, using the same frequency values. What change is there in the shape of the curve? Repeat the experiment using a 1 K resistor in place of the 10 K one, and plot all three graphs on the same scale.

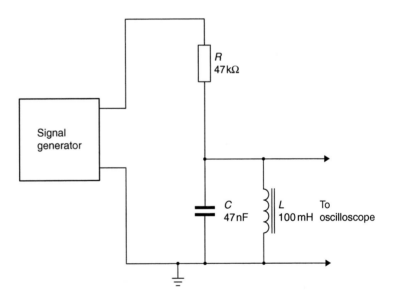

Figure 26.17   Circuit for practical exercise, parallel resonance

Bandwidth

The results of these experiments will show that the addition of either capacitance or inductance to a parallel resonant circuit causes the frequency of resonance to become lower. The addition of resistance in parallel has little effect on the frequency of resonance, but a considerable effect on the shape of the resonance curve. The effect of adding a small value of resistance is to lower the peak of the resonance curve, as might be expected because the sum of two resistors in parallel is a net resistance smaller than either. In addition, however, the width of the curve is increased.

A resistor used in this way is called a **damping resistor** or **de-Quing resistor**, since it reduces the Q of the circuit. Its effect is to make the resonant circuit respond to a wider range of frequencies, but at a lower amplitude. A damping resistor therefore increases the bandwidth of a resonant circuit, making the circuit less selective of frequency. When a parallel resonant circuit is used as the load of an amplifier, the tuned frequency is the resonant frequency of the parallel circuit, and the amount of damping resistance used will determine the bandwidth of the amplifier (Figure 26.18).

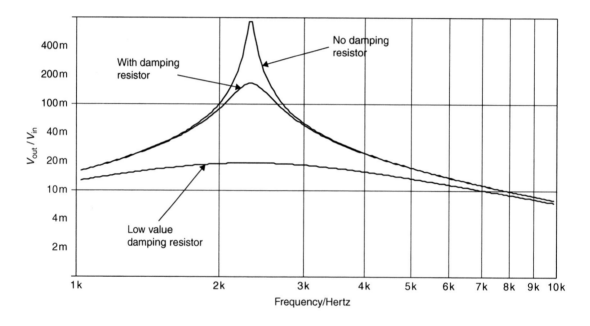

Figure 26.18   Effect of adding a damping resistor to a parallel resonant circuit

Using LCR circuits allows us to construct band-pass and band-stop filters which are more efficient than those described above. The simple LC parallel or series circuits can have resistors added to dampen the resonance so that the resonant effect is reduced but spread over a wider range of frequencies, providing a simple band-pass or band-stop action according to where the resonant circuit is placed.

For many purposes, however, these circuits do not provide a sharp enough distinction between the pass and the stop bands and much more elaborate filter circuits have to be devised. Band-pass and band-stop characteristics are often plotted on a linear scale of frequency so that the bandwidth is easier to read from the graph, but the attenuation is always plotted in terms of decibels.

Calculations on such filters are very difficult. Standard filter tables are published which reduce the design effort to scaling for the impedance and frequency, and computer programs such as SPICE can be used to graph the response for any combination of components. Figure 26.19(a) shows a BPF, using a combination of series and parallel circuit to determine the band-pass mid-frequency. The graph in Figure 26.19(b) shows the response of this circuit for the component values shown. It should be noted that there is significant insertion loss and ripple in the pass band. The ripple and losses can be minimized by careful design taking account of factors like the d.c. resistance of the inductors and tolerance of all the components, but this is beyond the scope of this book.

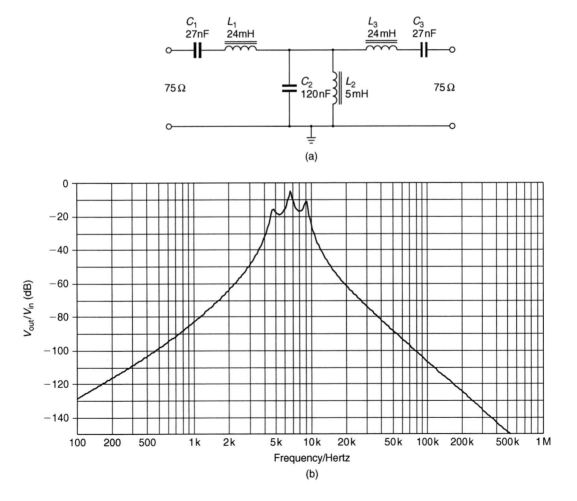

Figure 26.19   (a) Circuit, and (b) computed response curve for a band-pass filter

Multiple-choice revision questions

26.1 In a series circuit made from a resistor
and a capacitor the voltage across the
resistor is 100 V and the total circuit
voltage is 141.4 V. What is the voltage
across the capacitor?
(a) 50 V
(b) 75 V
(c) 100 V
(d) 150 V.

26.2 In the circuit in question 26.1 the current
is 1 A. What is the value of the resistor?
(a) 1 Ω
(b) 10 Ω
(c) 100 Ω
(d) 1000 Ω.

26.3 In the circuit in question 26.1 the current
is 1 A. What is the value of the capacitor
if the source frequency is 50 Hz?
(a) 16 μF
(b) 32 μF

(c) 64 μF
(d) 100 μF.

26.4 An RC low-pass filter uses a 47 kΩ
resistor and a 33 nF capacitor. What is the
frequency at which the output voltage is
−3 dB relative to the input?
(a) 97.4 kHz
(b) 645 kHz
(c) 103 kHz
(d) 155.1 kHz.

26.5 A parallel circuit consisting of a variable
capacitor and a 160 μH coil is used in a
radio receiver to select a station whose
frequency is 1215 kHz. What value will
the variable capacitor have when the
station is correctly tuned?
(a) 75.9 pF
(b) 107.2 pF
(c) 151.8 pF
(d) 300 pF.

27 Transformers and power transfer

A transformer consists of two or more inductors so wound that their magnetic fields interact. Usually the inductive coupling is maximized by winding the complete set of coils on a common closed **magnetic core** like a toroid (Figure 27.1b) or the typical mains transformer core which consists of either a stack of E-I or C-T shaped soft iron laminations (Figure 27.1a). **Toroidal** cores made from soft iron laminations are often used for mains transformers in equipment that is sensitive to stray magnetic fields, such as audio amplifiers. The windings are referred to as the **primary** (input) and **secondary** (output), respectively.

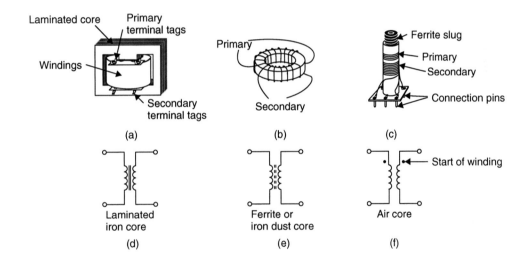

Figure 27.1   Transformer types and symbols

For audio applications or low-frequency switch mode power supply units, the cores may be made of either silicon iron alloy or mumetal. Mumetal is an alloy that consists mainly of nickel, iron, copper, manganese and chromium. For higher frequency applications, the core may be air, ferrite or powdered iron, also called iron dust.

Variable coupling between the primary and secondary of solenoid wound transformers (Figure 27.1c) can be achieved by position of a ferrite 'slug', which may be adjusted to tune the circuits to resonance, so avoiding the need for a variable capacitor. Most of the modern power supply units that operate on the switched mode principle use transformers that are wound on ferrite cores, which may be toroid, E-I or enclosed bobbin cores (often called pot cores).

Some typical construction methods for transformers are illustrated in Figure 27.1. Type (a) would be used for general-purpose mains and audio frequency work. A typical toroid for radio frequency (RF) use is shown in Figure 27.1(b) and a variable solenoid typical of an intermediate-frequency transformer in a radio receiver in Figure 27.1(c). The symbols used in circuit diagrams are shown in Figure 27.1(d–f). The type of core is indicated by the lines between the windings, while no lines indicates an air core.

The construction of typical general-purpose mains transformers is shown in Figure 27.2, the choice of lamination type E-I or C-T and the use of a split bobbin to separate the primary for the secondary being dependent on the exact application of the transformer.

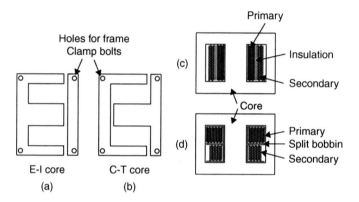

Figure 27.2 Construction of typical mains transformers: (a) E-I and (b) C-T core laminations, (c) secondary wound over primary, and (d) split bobbin winding

The inputs may be provided either by the mains power system or by other signals, which may be either pure a.c. or varying level unidirectional currents. These are applied to the primary winding and induce output currents, which are always a.c., across the terminals of the secondary winding(s). Since the mutual magnetic induction depends entirely on a changing input voltage level, a steady d.c. input current in the primary winding would induce zero output signal in the secondary windings. Although a transformer primary can carry a direct current it is important that any direct current is not large enough to cause the magnetic field to saturate the core of the transformer, since this would reduce efficiency and introduce distortion in the output signal. Some transformers have a gap in the core to reduce the possibility of direct current in the primary (or secondary) causing the core to saturate. These gaps are very small (less than 0.5 mm) and may be found in C-T core audio transformers and pot cores for switch mode power supplies, etc.

When the current is taken from the secondary winding(s) by connecting a load, an increased primary current must flow to provide the power that is being dissipated. If no current is taken from the secondary winding, the residual current flow in the primary winding, described as the magnetizing current, will be very small.

The input and output signal voltages may be in-phase or antiphase, depending on the polarity of the secondary connections and the relative directions of the windings. The secondaries may consist of a single winding, multiple separate windings or a single winding with multiple voltage taps each designed to provide a particular level of output voltage. Some secondary windings may be centre tapped to provide balanced output voltages.

The ideal transformer

The ideal transformer would be one that has no loss of power when in use, so that no primary current at all would flow until a secondary current was being drawn. Large transformers come quite close to this ideal, which is used for basic transformer calculations. In an ideal transformer,

$$\frac{V'_\mathrm{s}}{V'_\mathrm{p}} = \frac{n_\mathrm{s}}{n_\mathrm{p}}$$

where V'_s = secondary a.c. voltage, V'_p = primary a.c. voltage; n_s = number of turns of wire in the secondary winding, and n_p = number of turns of wire in primary winding.

> **Example:** A transformer has 7500 turns in its primary winding and is connected to a 250 V 50 Hz supply. What a.c. voltage will be developed across its secondary winding if the latter has 500 turns?

Solution: Substituting the data in the formula, $V'_\mathrm{s}/250 = 500/7500 = 1/15$, so that $V'_\mathrm{s} = 250/15 = 16.67$ or nearly 17 V. In practice, because no transformer is perfect, the output voltage would probably be somewhat less than 16 V.

In the ideal transformer, the power input to the primary winding must be equal to the power taken from the secondary winding, so that $V'_\mathrm{p} \times I'_\mathrm{p} = V'_\mathrm{s} \times I'_\mathrm{s}$ or, rearranging, $\frac{V'_\mathrm{s}}{V'_\mathrm{p}} = \frac{I'_\mathrm{p}}{I'_\mathrm{s}}$. Since $\frac{V'_\mathrm{s}}{V'_\mathrm{p}} = \frac{n_\mathrm{s}}{n_\mathrm{p}}$, it follows that $\frac{I'_\mathrm{p}}{I'_\mathrm{s}} = \frac{n_\mathrm{s}}{n_\mathrm{p}}$ or $I'_\mathrm{p} \times n_\mathrm{p} = I'_\mathrm{s} \times n_\mathrm{s}$

This last equation is often a convenient form which relates the signal currents in the perfect transformer to the number of turns in each of the two windings.

Transformer applications

Transformers are used in electrical circuits for the following purposes:

- voltage transformation: converting large signal voltages into low voltages, or vice versa, with practically no loss of power
- current transformation: converting low-current signals into high-current signals, or vice versa, with practically no loss of power
- impedance transformation: enabling signals from a high-impedance source to be coupled to a low impedance, or vice versa, with practically no loss of power through mismatch

- electrical isolation: for service purposes an isolating transformer, positioned between the mains supply and any equipment being worked on, will avoid electrical shocks to the operator.

Note that the transformer is a passive device without power gain. If a transformer has a voltage step-up of 10 times, it will also have a current step-down of 10 times (assuming no losses en route).

Example: The secondary winding of a transformer supplies 500 V at 1 A. What current is taken by the 250 V primary?

Solution: Since $V'_p \times I'_p = V'_s \times I'_s$, then $250 \times I'_p = 500 \times I'_s$, so that $I'_p = 2\,A$.

Transformers may also be used as matching devices so that the maximum power can be transferred from one circuit to another. The ideal method of delivering power to a load would be to use amplifiers that have a very low internal resistance, so that most of the power (I^2R) was dissipated in the load. Many audio amplifiers make use of such transistors to drive 8 ohm loudspeaker loads. For some purposes, however, transistors that have higher resistance must be used or loads that have very low resistance must be driven, and a transformer must be used to match the differing impedances.

In public address systems, for example, where loudspeakers are placed at considerable distances from the amplifier, it is normal to use 100 V line signals at low currents so as to avoid I^2R losses in the network. In such cases the loudspeakers are coupled to the lines through transformers.

From the voltage, current and turns ratios described above, we can deduce that the input and output load resistance values have the following relationship: $\dfrac{R_s}{R_p} = \left(\dfrac{n_s}{n_p}\right)^2$, which can be rearranged as $R_p = \left(\dfrac{n_p}{n_s}\right)^2 \times R_s$, so that the equivalent input resistance to signals entering a perfect transformer is $\left(\dfrac{n_p}{n_s}\right)^2 \times R_L$, where R_L is the load resistance connected to the secondary winding. The equivalent circuit for a perfect transformer is therefore that shown in Figure 27.3.

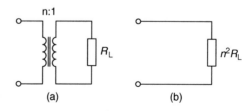

(a)   (b)

Figure 27.3   Equivalent circuit: (a) perfect transformer with load resistor, and (b) equivalent load resistor

For the maximum transfer of power, from an amplifier with output impedance Z_{OUT} to a load with resistance R_L the turns ratio $n = n_p/n_s$ can be expressed in the following way: $n = \sqrt{Z_{OUT}/R_L}$

Example: A power amplifier stage operates with a 64 ohm output impedance. What transformer ratio is needed for maximum power transfer to an 8 ohm load?

Solution: A 3:1 step-down transformer could therefore be used off the shelf, or a transformer specially wound for a 2.8:1 ($\sqrt{8}$) ratio.

Practical 27.1

Set up the circuit shown in Figure 27.4. Use a small mains trans-former, like a 12 V or 18 V output type and a 1 kΩ 2 W variable resis-tor or similar as the variable load resistor. Measure the load voltage and load current for different settings of the variable load resistor, and plot a graph of load power ($P = V.I$) on the y-axis against load current on the x-axis. What can you deduce from the graph about the trans-former ratio? Try swapping the primary and secondary connections and repeating the experiment. Explain the difference in the graph. What resistor settings give peak power transfer?

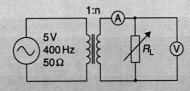

Figure 27.4   Circuit for Practical 27.1

Transformer power ratings

Owing to the nature of the self-inductance and capacitance of the transformer and the effects of the load, the a.c. voltage and current in the secondary circuit are rarely in phase. The power loading measured in watts can therefore be misleading and a more meaningful assessment uses the term **volt amps (VA)**. Furthermore, the rising voltage drop that occurs as the load current increases is described by the **regulation factor**. Typically, in small to medium-sized equipment power units, this accounts for losses of about 10%.

Consider the calculations associated with the following transformer designed to provide a secondary supply of 12 V at 4 A or 48 VA. This would produce an output of 12 V when driving a 48 W load. As the load is reduced, the voltage will rise owing to the regulation factor by about 10% to 13.2 V. (Note that this is not the same as the d.c. output after rectification).

Using a typical value of 4.8 turns per volt plus an extra 1% for each 10 VA of loading produces 12 × [4.8 +(1% of 4.8)] or approximately 60 secondary turns. Looking up wire tables would show that this loading could be safely supported by 1.25 mm diameter wire.

For the 250 V primary winding at 60 turns/12 V (or 5 turns/volt), this will require 250 × 5 = 1250 turns.

Maximum power transfer

If a generator and load are both resistive (*V* and *I* in phase), then the maximum transfer of power occurs when the internal generator resistance (R_G) and load resistance (R_L) are equal (i.e. when $R_G = R_L$). Matching of these values can be achieved using a transformer.

When the impedance of a generator or load has a reactive component (so that *V* and *I* are not in phase), maximum transfer of power occurs when the magnitudes of the impedances are equal, but with equal and opposite phase angles, i.e. when $Z(\Phi) = Z(-\Phi)$, where *Z* is the magnitude of the impedances and Φ is the phase angle. By using equal and opposite phase angles, the two parts of the circuit are brought into resonance to ensure the maximum transfer of power. This result explains why some industrial mains power inputs incorporate **power factor correction** capacitors.

Transformer losses

The three main types of power loss that occur in a transformer are:

- I^2R losses caused by the resistance of the windings
- eddy-current and stray inductance losses caused by unwanted magnetic interactions
- hysteresis loss arising from the core material (if a core is used).

Taking these in turn, I^2R (or joule) losses are those that are always incurred in any circuit when a current, steady or a.c., flows through a resistance. These losses can be reduced in a transformer by making the resistance of each winding as low as possible, consistent with the correct number of turns and the size of the transformer.

Joule losses are generally insignificant in small transformers used at radio frequencies, but they will cause overheating of mains transformers, particularly if more than the rated current is drawn or if ventilation is inadequate.

Stray inductance and eddy-current losses are often more serious. An ideal transformer would be constructed so that all the magnetic field of the primary circuit coupled perfectly into the secondary winding. Only toroidal (ring-shaped) transformers come close to this ideal. In practice, the primary winding generates a strong alternating field which is detectable at some distance from the transformer, causing a loss of energy by what is termed stray inductance.

In addition, the alternating field of the primary can cause stray voltages to be induced in any conducting material used in the core or casing of the transformer, so that unwanted currents, called eddy currents, flow. Since additional primary current must flow to sustain these eddy currents, they cause a loss of power, which can be significant.

The problem of eddy currents in the core is tackled in two ways:

- The core is constructed of thin laminations clamped together, with an insulating film coating on each to lessen or eliminate conductivity.
- The core is constructed from a material that has high resistivity, such as ferrite.

The third type of loss, called **hysteresis** loss, occurs only when a magnetic core is used. It represents the quantity of energy that is lost when a material is magnetized and demagnetized. This type of loss can be minimized only by careful choice of the core size and material for any particular transformer.

Hysteresis losses will, however, increase greatly if the magnetic properties of the core material change, or if the material becomes magnetically saturated. The following precautions should therefore be taken in connection with transformers:

- Do not dismantle transformer cores unnecessarily, or loosen their clamping screws.
- Never bring strong magnets near to a transformer core.
- Never pass d.c. through a transformer winding unless the rated value of the d.c. is known and is checked to be correct.

Types of transformer

Mains supply

Mains frequency is low and fixed at either 50 or 60 Hz. A substantial core is required which must be laminated (hysteresis losses can be reduced to negligible proportions by careful choice of a core material). Where an external magnetic field is especially undesirable (as in audio amplifiers and cathode ray oscilloscopes), a toroidal core can be used with advantage.

Figure 27.5(c) shows a mains-type transformer with split primary windings. This set-up allows the transformer to be switched from 230 V input, using the two windings connected in series, to 120 V operation, using the two windings connected in parallel. Many items of equipment that have

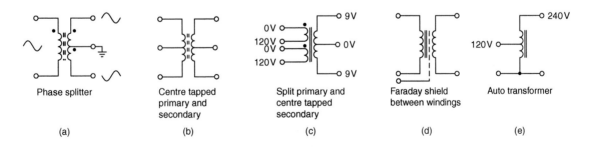

Figure 27.5 Types of transformer winding: (a) phase splitter, (b) centre tapped windings, (c) split primary mains transformer, (d) Faraday shield to minimize capacitive coupling between windings, and (e) step-up autotransformer

a mains selector switch to switch from 120 V to 240 V input use this type of transformer. Industrial equipment that is manufactured for both the US and European market often uses just one type of transformer, with the links being hardwired at assembly time and replacement power supply units often being shipped unconfigured. The increasing use of universal input switch mode power supplies has made this much less common than it was.

Audio frequency

The core material must be chosen from materials causing only low hysteresis loss because of the higher frequencies that will be encountered. The windings must be arranged so that stray capacitance between turns is minimized. In general, any flow of d.c. is undesirable.

Medium- to high-frequency RF and VHF

In this range, the losses from laminated cores are unacceptably high, so that iron dust, ferrite or air cores must be used. Because of the high frequencies involved, a small number of turns is sufficient for each winding. Stray fields are difficult to control, so that screening (see below) is often needed.

UHF and microwave

Only air cores and specialized ferrite materials can be used in this range, and 'coils' may actually consist of less than one full turn of wire. They may even consist of short lengths of parallel wire. Unwanted coupling becomes a major problem, so that the physical layout of components near the transformer assumes great importance.

Standard windings

Figure 27.5(a) shows how a transformer with a centre tapped secondary may be used to provide antiphase outputs, such as required for driving power output stages of linear amplifiers or for full wave rectification purposes.

The autotransformer is a single-tapped winding, shared by both input and output, equivalent to the use of a double-wound transformer with a common primary and secondary terminal. The ratio of input/output voltages and currents still follows the normal transformer relationships. An autotransformer with a variable tapping position (such as the Variac™) is used for providing variable voltage a.c. supplies. Note, however, that such transformers provide no isolation between their primary and secondary windings (Figure 27.5e).

Bifilar winding, twisting the primary and secondary wire together before winding them onto the core, is a method of providing very close coupling between primary and secondary windings, it is particularly useful in audio and radio frequency transformers. If a tapped primary or secondary is required then this method of construction can be extended to three or more cores (trifilar, etc.). Because the primary and the secondary turns are wound together, rather than in separate layers, there is significant interwinding capacitance.

Components such as printed circuit board tracks, wiring and inductors may need to be shielded from the magnetic field created by transformers. This is difficult; it is much better to use a core design that minimizes the external field, such as the toroid.

Electromagnetic screening requires the use of high-permeability alloys such as mumetal or super-permalloy to encase the device to be protected.

Boxing a component in such a way ensures that any magnetic fields are contained within the casing, preventing the fields from entering the enclosed space.

Electrostatic screening is comparatively easy, since any earthed metal between a component and the transformer will screen the component from the electrostatic field of a transformer. When interaction between windings needs to be prevented a Faraday shield is placed between the windings, usually consisting of a copper foil strip the width of the winding on the transformer. This must be insulated so that it does not make a complete conducting turn around the core, since this would be a shorted turn, causing the output of the secondary to be greatly reduced and a very high current to flow in the primary, possibly leading to overheating and failure (Figure 27.5d).

Practical 27.2

Using a transformer of known turns ratio, preferably a type using a tapped secondary winding, connect the circuit of Figure 27.6. An effective component to use in this experiment is a toroidal core with a 240 V primary winding, obtainable from most educational suppliers. Measure the a.c. input and output voltages for each set of taps, and find the values of V_s/V_p. Compare these values with the known values of the turns ratio. The secondary load resistor is optional, but if used should be around 1 kΩ and helps by ensuring that some secondary current flows.

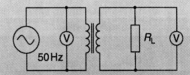

Figure 27.6   Arrangement for Practical 27.2

The following are common transformer faults, with hints on how to detect and remedy them:

- Open-circuit windings can be detected by ohmmeter tests. A winding may also acquire high resistance, which is caused by high-resistance internal connections, typically 100 K instead of 100 ohms.

- Short-circuit turns: these are difficult to detect because the change of resistance is very small. Even a single turn that is short-circuited will dissipated considerable energy while making practically no difference to the d.c. resistance, making it very difficult to locate. Shorted turns will cause an abnormally large primary current to flow even when the secondary is disconnected, so that mains transformers overheat and transformers operating at high frequencies fail completely. This is a fault that particularly affects television line output transformers. The simplest test and cure is replacement by a component known to be good.

- Loose, damaged or missing cores: loose cores will cause mains transformers to buzz and overheat. Cracked or absent ferrite cores in radio-frequency transformers will cause mistuning of the stage in which the fault occurs.

Thermal protection of transformers

Power supply circuits almost invariably contain protection devices, to provide overvoltage and overcurrent protection for the equipment being powered. Supplies and transformers also require protection and there are also the requirements of product safety, to prevent overload or fault conditions resulting in fire or electric shock hazard to the user.

Fuses

Fuses or fuse links are the most commonly used safety critical device for protection against overload conditions, but they are also open to misuse. The fuse link should always be replaced with one of the correct rating after the fault condition has been cleared.

The fuse-link rating is the current that the device will carry for a long period. It will often carry a current that is 25% in excess of the rated value for 1 h or more. A further parameter is the joule rating, which is given by I^2t, showing that the failure depends on the square of the current and the time for which it is flowing. The minimum fusing current is typically 50–100% above the rated value. Hence, many fuse links are described as slow-blow devices.

Fuse links usually have a voltage rating that is different for a.c. and d.c. applications.

Embedded devices

These devices are often buried within the winding space of a transformer and wired in series with the primary winding. When the current exceeds some value the temperature of the transformer rises, increasing the resistance of the thermistor so that the input current falls to lower the temperature.

An embedded bimetallic operated switch with its contact wired in series with the primary current has been used in the past. The switch contacts open when the transformer temperature rises above some predetermined level to provide protection. When the temperature falls this system is self-resetting.

Another embedded protection device found in low-cost transformers such as those used in mobile phone chargers consists of a conducting spring strip soldered at one end to a contact with low melting-point solder. If the temperature exceeds the melting point of the solder the strip springs away from the contact, breaking the circuit. These devices cannot usually be reset.

Note: for all resettable embedded devices, a continuous cyclic switching action indicates a fault in urgent need of attention.

Multiple-choice revision questions

27.1 A step-down transformer uses a 9:1 ratio. If the primary voltage is 240 V a.c., what (assuming no losses) is the secondary output?
  (a) 53.3 V
  (b) 36 V
  (c) 26.7 V
  (d) 25 V.

27.2 A transformer operates from 240 V mains and delivers an output of 60 W at 6 V. Assuming no losses, what amount of primary current flows?
  (a) 0.1 A
  (b) 0.25 A
  (c) 2.5 A
  (d) 10 A.

27.3 A transformer is to be used to match a source resistance of 100 ohms to a loudspeaker of 4 ohms. What ratio is needed?
  (a) 25:1
  (b) 16:1
  (c) 5:1
  (d) 4:1.

27.4 A 470 Ω resistor is connected across the secondary of an ideal transformer with a 20:1 turns ratio and the primary is connected to the mains supply at 230 V r. m.s. How much power is dissipated in the resistor?
  (a) 127 mW
  (b) 281 mW
  (c) 489 mW
  (d) 5.63 W.

27.5 A transformer with a prewound primary is required to provide a low-current 33 V output. If the mains input is 230 V and the primary winding has 1000 turns, how many secondary turns are needed?
  (a) 121
  (b) 143
  (c) 182
  (d) 286.

28 Semiconductors, active devices and transducers

The majority of semiconductor devices are made from silicon (Si), including most transistors, diodes and integrated circuits (ICs). Applications of other semiconductor materials are mostly in specialized areas such as light-emitting diodes (LEDs), where compound materials such as aluminium gallium arsenide (AlGaAs) produce red, yellow and green light, and silicon nitride (SiN) blue light. Some very old equipment will still have germanium transistors and diodes, but with the exception of germanium in point contact detector diodes such as the OA47 these have not be used in new designs for more than 20 years. Gallium arsenide finds applications in microwave transistors and very fast logic used in long-haul fibreoptic telecom and supercomputing applications. A relatively recent development is the use of silicon on a germanium substrate for the fabrication of very fast devices.

Silicon, which is found as silicon oxide (sand), is the second most abundant element on Earth; however, the exceptionally high purity of the material needed to make ICs and transistors makes refined silicon very expensive.

Pure silicon on its own does not make a transistor or diode; rather, the silicon has to have exact small quantities of other elements such as phosphorus (P) or boron (B) diffused into the crystal lattice. The resulting silicon is said to be doped with donor or acceptor atoms; this is because they provide either extra electrons or extra spaces (holes) for electrons, compared with what the pure silicon would have provided. Phosphorus has five valence electrons compared with silicon's four, so it is a donor of electrons, and material doped with phosphorus is said to be N-type. Boron has three valence electrons, so it is an acceptor atom, and silicon doped with it is called P-type. If two regions of different doping are formed so that they have a common boundary it is called a PN junction. The PN junction is the basis of diodes and bipolar transistors, which are referred to as NPN or PNP depending on the junctions that they contain.

Current flowing from the P-type region to the N-type region is called the forward current. The PN junction has an inbuilt potential or threshold voltage below which it will not conduct a forward current; only once an externally applied voltage has overcome the threshold voltage will the diode begin to conduct (Figure 28.1). In the reverse direction the diode does not conduct until a breakdown threshold is reached, which is usually 20 or more times the forward threshold.

Although germanium has a lower inherent threshold voltage (100 mV) than silicon, it now finds relatively few applications in modern semiconductor

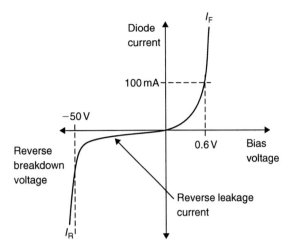

Figure 28.1   Typical silicon diode characteristic

technology. For both germanium and silicon, the reverse biased leakage current of a PN junction approximately doubles for every 11°C rise in temperature. Since germanium has a significantly higher leakage at room temperature, its operating range is restricted to about 75°C maximum. This compares unfavourably with silicon, which can operate effectively up to around 250°C.

The PN junction threshold voltage for silicon is about 600 mV at room temperature. It falls by about 2 mV per degree Celsius as the temperature rises, so a forward-biased diode can be used a temperature sensor. For example, a 1N4148 diode changes from 750 mV at –55°C to 350 mV at +150°C when biased with a current of 200 μA. This can be achieved with a 22 kΩ resistor and a 5 V supply.

Bipolar junction transistor

A transistor has three terminals and therefore two PN junctions. The collector and emitter are the connections to the outer layers, while the base is the middle layer. The base is a very thin layer and its function is to control the flow of current between the collector and emitter. The base emitter junction is forward biased to allow current flow between the collector and the emitter; in operation, the collector base junction is usually reverse biased. Figure 28.2 shows the symbols for NPN and PNP transistors.

Tests with an ohmmeter can identify bipolar junction transistor (BJT) junction faults. A good transistor should have a very high resistance reading between collector and emitter with either polarity of connection. Measurements between the base and either of the other two electrodes should show one conducting direction and one non-conducting direction. Any variation from this pattern indicates a faulty transistor with either an open-circuit junction (no conduction in either direction) or excessive leakage (conduction in both directions).

The current flowing between the collector and the emitter of a bipolar transistor is much greater than that flowing between the base and the emitter,

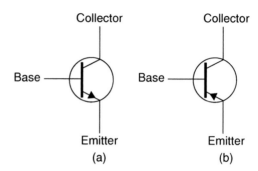

Figure 28.2   Schematic symbols: (a) NPN, and (b) PNP transistors

and the collector current is controlled by the base current. The ratio of collector current to base current is constant (given a constant collector-to-emitter voltage) and is commonly called the current gain for the transistor (its full name is the common-emitter current gain). The symbol used to indicate it is h_{fe}. A low-gain transistor may have a value of h_{fe} of around 20–50, a high-gain transistor one of 300–800 or more.

> Note that the tolerance of values of h_{fe} is very large, so that transistors of the same type, even transistors coming from the same batch, may have widely different h_{fe} values. Manufacturers often use a gain group code after the part number to make variations within groups easier to deal with; for example, the BC547, which has an h_{fe} range from about 80 to 400, is available in gain groups as BC547A h_{fe} 110–220, BC547B h_{fe} 200–450 and BC547C h_{fe} 420–800.

When one transistor is substituted one for another, the following rules should be obeyed:

* The substitute transistor must be of the same type (i.e. silicon, NPN, switching as opposed to amplifying, etc.).
* The substitute transistor should have about the same h_{fe} value.
* The substitute transistor should have the same ratings of maximum voltage and current.
* When making such substitutions, it is not always possible to guarantee the in-circuit performance of the change.

Bipolar transistors are used as current amplifiers, voltage amplifiers, oscillators and switches. An amplifier has two input and two output terminals, but a transistor has only three electrodes. It can therefore only operate as an amplifier if one of its three electrodes is made common to both input and output circuits.

Any one of a transistor's three electrodes can be connected to perform in this common role, so there are three possible configurations: common emitter, common collector and common base. The three types of connection are shown in Figure 28.3(a–c).

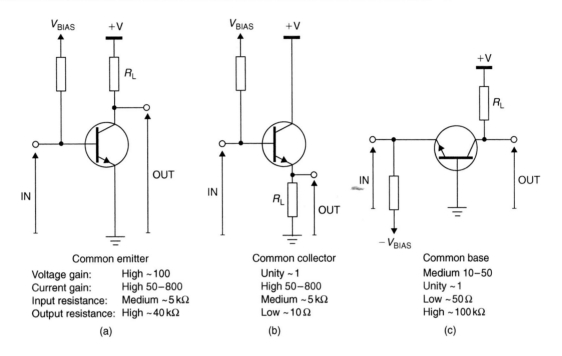

Figure 28.3   The three circuit connections of a bipolar transistor

The normal function of a transistor when the base-emitter junction is forward biased and the base-collector junction is reverse biased, is as a current amplifier. Voltage amplification is achieved by connecting a load resistor (or impedance) between the collector lead and the supply voltage (Figure 28.3a). Oscillation is achieved when the transistor is connected as an amplifier with its output fed back, in phase, to its input. The transistor can also be used as a switch when the small base-emitter junction current is used to switch on the larger collector current.

The three basic bipolar transistor circuit connections are shown in Figure 28.3, with applications and values of typical input and output resistances given below each. The common-collector connection in Figure 28.3(b), with signal into the base and out from the emitter, is used for matching impedances, since it has a high input impedance and a low output impedance. The common-base connection, with signal into the emitter and out from the collector (Figure 28.3c) is often used for ultra high-frequency (UHF) amplification, for example in television mast head amplifiers.

The graph of a typical transistor I_C versus V_{CE} for different values of base current (Figure 28.4) shows how the base current can be used to control the transistor collector current; for example, changes in the collector voltage can be seen to have only relatively small effects on the collector current, implying a high output resistance.

Transistor failure

When transistors fail, the fault is either a short-circuit (s/c) or an open-circuit (o/c) junction; or the failure may possibly be in both junctions at the same

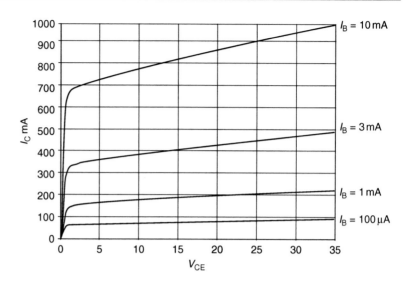

Figure 28.4   Typical small signal NPN transistor characteristic

time. An o/c base-emitter junction makes the transistor 'dead', with no current flowing in either the base or the collector circuits.

When a base-emitter junction goes o/c, the voltage between the base and emitter may rise higher than the normal 0.6 V (silicon) or 0.2 V (germanium), although higher voltage readings are common on fully operational power transistors when large currents are flowing. An s/c base-emitter junction will allow current to flow easily between these terminals with no voltage drop, but with no current flowing in the collector circuit.

The two above faults are by far the most common, but sometimes a base-collector junction goes s/c, causing current to flow uncontrollably. The base region of a bipolar device can be ruptured through the application of an excessively high collector voltage, this is often described as punch-through.

All of these faults can be found by voltage readings in a circuit, or by use of the ohmmeter or transistor tester when the transistor is removed from the circuit. (Transistor testers are now available that allow in-circuit testing.)

Field-effect transistors

The BJT relies for its action on making a reverse-biased junction conductive by injecting current carriers (electrons or holes) into it from the other junction. The principles of the field-effect transistor (FET) are entirely different. In any type of FET, a strip of semiconductor material of one type (P or N) is made either more or less conductive because of the presence of an electric field pushing carriers into the semiconductor or pulling them away.

There are two types of FET: the junction field-effect transistor (JFET) and the metal oxide semiconductor field-effect transistor (MOSFET). Both work by controlling the flow of current carriers in a narrow channel of semiconductor, usually silicon. The main difference between them lies in the method used to control the flow. Since the JFET is now a device that is rarely used in consumer electronics except for high-end audio preamplifiers, we shall concentrate on the much more common MOSFET.

Figure 28.5(a) shows the basic construction of a MOSFET. A silicon layer, called the substrate or base or sometimes bulk, is used as a foundation on which the FET is constructed. The substrate is lightly doped and may have a separate electrical connection, but it generally takes no part in the FET action and if a separate electrical connection is provided it is usually connected to the source, forming a reverse-biased diode between source and drain under normal operating conditions. Two heavily doped regions, of opposite polarity to the substrate, forming the source and drain, are then diffused into the top surface of the substrate, and an insulating layer of silicon dioxide (SiO_2) is grown over the surface. In the example, the substrate is of P-type silicon, and the source and drain are of N-type, so there is no conducting path between the N-type source and drain regions.

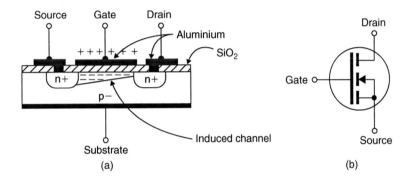

Figure 28.5 MOSFET principles, N-channel device: (a) basic construction, and (b) schematic symbol

The gate is insulated from the channel by a thin film of silicon dioxide, obtained by oxidizing the surface of the silicon, and a metal film is deposited over this insulating layer to form the gate. Before deposition the metal film holes are etched in the SiO_2 above the drain and source diffusions so that aluminium contacts can be made to the diffused regions.

For an N-channel MOSFET with a gate-to-source voltage equal to zero, the device is cut off. When a gate voltage that is positive with respect to the substrate is applied, an electric field is set up that attracts electrons towards the oxide layer. These now form an induced channel to support a current flow. An increase in this positive gate voltage will cause the drain-to-source current flow to rise (Figure 28.6). A negative voltage so applied would repel any electrons from the channel and so reduce its conductivity.

If a positive voltage is applied to the gate of an N-channel device, it has the effect of attracting electrons into the region between the source and drain under the gate electrode. This is the same effect as observed in charging a capacitor; that is, current flows into the gate of the MOSFET until the gate capacitance is fully charged. This charging current can be quite high, and the rate at which current can be moved into and out of the gate limits the speed at which the device can be turned on or off.

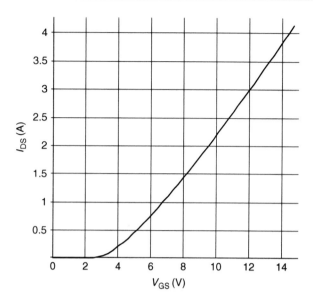

Figure 28.6   Drain current dependence on gate voltage for typical medium-power N-channel MOSFET

Both N-channel and P-channel devices can be made. In addition, the channel can be either lightly or heavily doped with the same polarity as the source and drain. In this case there will be a conducting path of fairly low resistance between the source and the drain when no bias is applied to the gate. Such a device is usually operated with a bias on the gate that will reduce the source-drain current, and is said to be used in **depletion mode**. When the channel is formed from lightly doped or substrate material it is normally non-conducting, and its conductivity is increased by applying bias to the gate in the correct polarity, using the FET in **enhancement mode**. Enhancement-mode operation is more common.

Small-signal MOSFETs need to be handled with great care because the gate is completely insulated from the other two electrodes by the thin film of silicon oxide. This insulation will break down at a voltage of 20–100 V, depending on the thickness of the oxide film. When it does break down, the transistor is destroyed. Any insulating material that has rubbed against another material can carry voltages of many thousands of volts, and lesser electrostatic voltages are often present on human fingers. There is also the danger of induced voltages from the a.c. mains supply.

Voltages of this type cause no damage to bipolar transistors because these devices have enough leakage resistance to discharge the voltage harmlessly. The high resistance of the MOSFET gate, however, ensures that electrostatic voltages cannot be discharged in this way, so that damage to the gate of a MOSFET is always possible.

To avoid such damage, most MOSFET gates that are connected to external pins of ICs of power MOSFETs are protected by diodes, which are created as part of the IC or FET during manufacture and have a relatively low reverse breakdown voltage. These protecting diodes will conduct if a voltage at a gate terminal becomes too high or too low compared to the source or drain voltage level, so avoiding breakdown of the insulation of the gate.

The use of protective diodes makes the risk of electrostatic damage very slight for modern MOS devices, and there is never any risk of damage to a gate that is connected through a resistor to a source or drain unless excessive d.c. or signal voltages are applied. Nevertheless, it is advisable to take precautions against electrostatic damage, particularly in dry conditions and in places where artificial fibres and plastics are used extensively. These precautions are:

- Always keep new MOSFETs with conductive plastic foam wrapped round their leads until after they have been soldered in place.
- Always short the leads of a MOSFET together before unsoldering it.
- Never touch MOSFET leads with your fingers.
- Never plug a MOSFET into a holder when the circuit is switched on.

MOSFETs can be used in circuits similar to those in which bipolar transistors are used, but they give lower voltage gain and are only used singly when their peculiar advantages are required. They are extensively used in ICs, however, because of their low power dissipation and the ease of forming very large numbers of MOSFETs on a silicon chip. In circuit applications, MOSFETs have the following advantages:

- MOSFETs have a very high input resistance at the gate, a useful feature in voltmeter amplifiers.
- MOSFETs perform very well as switches, with channel resistance switching between a few hundred milliohms and several megaohms as gate voltage is varied.
- The graph of channel current I_{ds}, plotted against V_{gs}, the voltage between the gate and the source, is noticeably curved in a shape called a square law. This type of characteristic is particularly useful for signal mixers in superheterodyne receivers.

Double-gate MOSFETs are used as mixers and as radio frequency (RF) amplifiers in frequency modulation (FM) receivers. The shape of MOS characteristics can provide less distortion in linear power amplifiers, and high-power FETs are available for use in high-quality audio equipment.

MOSFET failure is almost always caused by breakdown of the insulating silicon oxide layer. In either case, gate voltage can no longer control current flow in the channel between source and drain, and pinch-off (the cutoff condition) becomes impossible. In addition, if very large currents have been allowed to flow between source and drain, the channel may overheat and rupture.

Four-layer devices

This topic covers a range of related semiconductor devices that are basically four-layer PNPN sandwiches designed either for a.c. or d.c. switching purposes.

The **thyristor** is a commonly used switching device, and often referred to as a silicon-controlled rectifier (**SCR**). It is fabricated as four doped layers, and its symbol and schematic construction are shown in Figure 28.7. Only three of the four semiconductor layers have external connections.

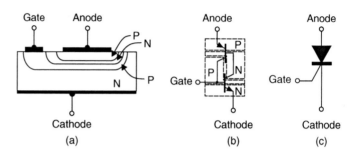

Figure 28.7   Silicon-controlled rectifier: (a) typical arrangement of layers, (b) equivalent circuit, and (c) symbol

When the gate voltage is zero, the device offers high resistance irrespective of the polarity of the anode-to-cathode voltage, because at least one PN junction is always reverse biased. Even when a thyristor is forward biased, with the anode positive with respect to the cathode, it will only be driven into conduction when its gate is forward biased sufficiently.

With the anode voltage positive with respect to the cathode, the action is controlled by the gate voltage. Briefly applying a positive voltage (V_g) to the gate causes the thyristor to go into conduction (between anode and cathode) and then remain in this state even if V_g is removed. Turn-off is achieved by briefly reducing the anode current below some threshold value, typically a few milliamps. The device can carry several tens of amps at several hundreds of volts. Referring to the equivalent circuit (Figure 28.7b), we can see that once the gate (base of the NPN transistor) is turned on the device latches because the gate current is supplied by the top (PNP) transistor's collector. From this, it can also be deduced that to turn off the device the gate current must be reduced below the holding threshold.

Once it has been switched on in this way, the thyristor will remain conducting until either:

• the voltage between anode and cathode falls to a small fraction of a volt; or

• the current flow between anode and cathode falls to a very low value.

The important ratings for a thyristor are its maximum average current flow, its peak inverse voltage, and its values of gate-firing voltage and current. A small thyristor will fire at a very small value of gate current, but a large one calls for considerably greater firing current (1 A or more).

The silicon-controlled switch (**SCS**) is constructed in a very similar way to the SCR, is self-latching and functions in a similar way for turn-on, but is turned off by either reducing the anode current or reverse biasing the gate to cathode junction.

Practical 28.1

Connect up the circuit of Figure 28.8. Any small thyristor such as 2N5060 or TIC106D will be suitable. The purpose of the lamp bulb is to indicate when the thyristor has switched on, and limit the current once it has turned on. Use a d.c. supply as shown, increase the gate voltage from zero, measure the gate voltage and load current until the thyristor fires. After the lamp lights, disconnect the gate circuit, and note that this has no effect. Switch off the power supply, and then switch on again. The lamp will not now light because the thyristor has switched off. Repeat the operation, this time using an unsmoothed rectified a.c. supply. The thyristor now switches off when the gate voltage is switched off, because the unsmoothed supply reaches zero voltage 100 times per second (assuming a full-wave rectified 50 Hz supply).

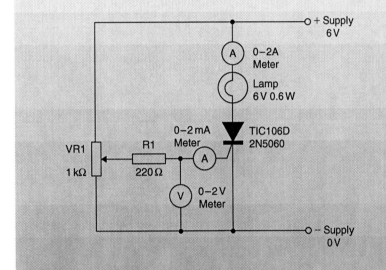

Figure 28.8   Circuit for practical

Thyristors are used for power control, using either a.c. or an unsmoothed rectified supply. By being made to fire at different points in each a.c. cycle, the thyristor can be made to conduct for different percentages of the cycle, thus controlling the average current flow through the load. This is called **phase control** of the duty cycle.

An alternative method used for slow-changing load conditions, such as in furnace temperature control, fires the thyristor into conduction for several a.c. cycles and then drives into cut-off for several further cycles. The average power in the load can be controlled by altering the ratio of conducting cycles to non-conducting cycles. This technique is known as **burst control**.

The gate turn-off thyristor (**GTO**) is an alternative device which is triggered on by a positive pulse of fairly high current (typically 100 mA) and triggered off by a negative pulse of lower current, both pulses being applied to the same gate.

The tetrode thyristor is equipped with both anode and cathode gates to provide a self-latching action. The turn-on action is achieved by driving the cathode gate positive or the anode gate negative, and the device is turned off by reversing the gate potentials.

Practical 28.2

Connect up the thyristor circuit illustrated in Figure 28.9, and switch on. Use the potentiometer to control lamp brightness by altering the time in the cycle at which the thyristor fires. Note that an unsmoothed full-wave rectified supply is essential.

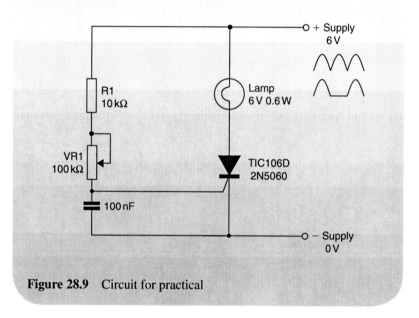

Figure 28.9   Circuit for practical

Failure of a thyristor can be caused by an open-circuit gate, or by internal short-circuits. When the gate is o/c, the thyristor will fail to conduct at any gate voltage. When an internal short-circuit is present, the thyristor acts like a diode, conducting whenever the anode is more than 0.6 V positive with respect to the cathode, so making control impossible. A completely short-circuit thyristor is able to conduct in either direction, with similar total loss of control.

Thyristor circuits often include a low-pass filter (called a **snubber** circuit) consisting of an inductor and capacitor or resistor and capacitor in the anode circuit, to suppress the RF interference caused by the sudden switch-on action of the thyristor.

A **triac** is a two-way thyristor which, when triggered, will conduct in either direction. The terminals are labelled M1, M2 and Gate (the words 'anode' and 'cathode' cannot be used in this case because the current can flow in either direction). The gate can be triggered by a pulse of either polarity, but the most reliable triggering is achieved when the gate is pulsed positive with respect to M1.

In most triac circuits, M1 and M2 have alternating voltages applied to them, so that the gate must receive the same waveform as M1 when it is not being triggered. To trigger the gate, a pulse must be added to the waveform already present, and this is most easily done by using a pulse transformer driven by a trigger circuit. The transformer being used to isolate the triac works at line voltage, from the low-voltage control circuit. Figure 28.10(a) shows the symbol for a triac.

A **diac** is a two-terminal trigger device fabricated as a gateless triac that triggers by breakdown at a designed low voltage and is often used in the gate circuit of a thyristor or triac. Figure 28.10(b) shows the schematic symbol for the diac. At low voltages or either polarity, a diac is completely non-conducting; however, once its breakdown threshold is reached it can conduct in either direction.

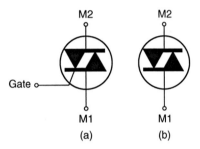

Figure 28.10   (a) Triac symbol, and (b) diac symbol

When a diac is used, the gate of the thyristor can be connected to the cathode by a low-value resistor to avoid accidental triggering. In addition, if the triggering waveform rises slowly, the diac will ensure that the thyristor is switched on by a fast rising pulse of current, so avoiding uncertain triggering times.

- The causes of failure in triacs are the same as they are for thyristors.

The advantages of using thyristors and triacs for power control are:

- Both thyristor and triac are switched either completely off, with no current flowing, or fully on, with only a small voltage between the terminals. Either way, very little power is dissipated in the semiconductor, so that heat sinks are required only for devices that handle a substantial amount of power.

- Operation can be either at line or at higher frequencies, unlike relays or similar electromechanical switches.

- Thyristors and triacs are 'self-latching', which means that they stay conducting once they have been triggered. A relay, by contrast, needs a current passed continuously through its coil to keep it switched over, and relay latching circuits require additional relay contacts.

Thyristors and triacs are used as solid-state relays, allowing a small voltage and current to switch a much larger voltage and current. Unlike a relay, however, there is no electrical isolation between the trigger circuit and the main supply unless a suitable transformer is used, so that some applications call specifically for a relay to be used in place of any semiconductor device.

Figure 28.11 shows a typical circuit using a triac to control the power to a 60 W lamp. The potentiometer adjusts the charging current to the capacitor, so determining at what part in each half-wave the voltage across the diac will be enough to cause conduction and so fire the triac.

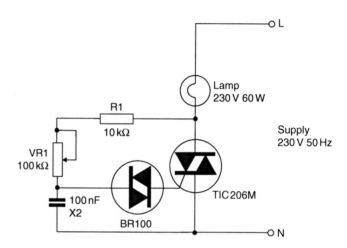

Figure 28.11   Typical triac and diac use in a mains lamp dimmer

HAZARD – high voltages: triac lamp dimmers like that of Figure 28.11 are connected to mains potential; when working on such circuits, the circuit must be supplied from an isolating transformer, and all live parts of the circuit should be covered to protect against accidental contact.

Transducers

A transducer is a device that converts energy from one form to another. The transducers that are of most interest as far as electronics is concerned are those in which one of the energy forms is electrical energy. Of the many possible transducers, the most common are the transducers for light, heat and sound. The transducers that involve light are known as electron-optical devices.

Photoelectric transducers convert light energy into electrical signals and are widely used in the communications and measurement environments. They are used extensively in position sensing and counting devices where the interruption of a light beam is converted into an electrical signal. As most of these devices have a wavelength response greater than the human eye, their applications extend into the infrared region. Two main principles can be used: photoconductivity and photovoltaic effect.

Light-dependent resistors

The resistance of certain materials that are used for light-sensitive devices decreases when subjected to increased illumination. This is a photoconductive effect. The common materials used for **light-dependent resistors** (**LDRs**) include selenium, cadmium sulphide, cadmium selenide and lead sulphide. Each material responds to different wavelengths of light. Figure 28.12(a) shows the general appearance of a photoconductive cell and its construction. Gold-conducting sections are deposited on a glass plate with a long, meandering gap to isolate the two sections. A thin layer of suitable photoconductive material is then deposited to bridge the insulating region. This construction is necessary to reduce the cell's resistance to a usable value.

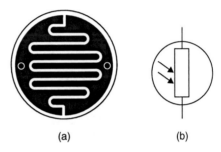

(a)   (b)

Figure 28.12   (a) Typical photoconductive cell, and (b) schematic symbol

Typical resistance values for various cells range from about 500 K to 10 M in darkness and from about 1 K to 100 K in bright light. The change in resistance is non-linear and there is a significant time lag (up to 0.1 s) in response to a pulse of light.

Typically, LDRs are used in series with a fixed resistor or potentiometer, to form a potential divider whose ratio is dependent on the light level. Increasing the illumination of the LDR lowers its resistance so that the output voltage changes. The maximum permitted voltage for these cells may be as high as 300 V and the maximum power dissipation is in the order of 300 mW. The dark-to-light resistance ratio ranges from about 50:1 to 250:1. These cells can often be used directly without amplification, to trigger a triac in applications such as dusk-to-dawn switches.

The cells are particularly sensitive to red light and the infrared wavelengths, so they are often used as flame detectors in boiler and furnace control systems. Application of an excessive voltage or current is the most common cause of failure.

Photodiodes

Photodiodes are very similar to a normal PN junction device. They are formed with one very thin region and equipped with a lens so that light energy can be directed into the depletion region. When this happens, hole–electron pairs are generated to increase the diode's conductivity. Such diodes have a peak response in the infrared region, but the response to visible light is still very useful.

The diodes are operated either reverse biased, or only very slightly forward biased (to increase sensitivity) so that no current flows. Typical dark current is as low as 2 nA, rising to about 100 μA in bright light for germanium types. Modern PIN silicon [a PN junction with an additional interposed layer of intrinsic (I)-type, meaning undoped] semiconductors may have a peak power dissipation greater than 100 mW.

The increase in current due to the light is practically linear. The response time to a pulse of light is very short, so they find applications in high-speed switching circuits. In general, the small output is a disadvantage and amplification is needed. The basic circuit configuration is shown in Figure 28.13.

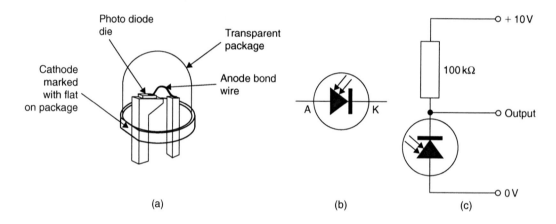

Figure 28.13   Photodiode: (a) package, (b) symbol, and (c) typical circuit

Phototransistors

A phototransistor has a similar construction to a silicon planar transistor, except that it is equipped with a lens so that light can be made to shine directly into the base-emitter junction. Their response time to a pulse of light is in the order of 1–2 μs and they are therefore mostly used in medium-speed light detector circuits. The current ranges from about 25–50 μA in darkness to about 5–10 mA in bright light.

The phototransistor (Figure 28.14) is connected so that the base is open-circuit, reversed biased or only very slightly forward biased. In some devices, the base connection is omitted altogether, so only the collector and emitter connections can be used. The load may be placed in either the collector or emitter leads as shown (left). Light shining into the base-emitter junction modulates the base current, which is in turn magnified by the current gain. Such a device can be used to drive a load directly, as shown in the relay-load example.

Light-emitting diodes

The operation of the photodiode depends on the application of energy to generate hole–electron pairs. The reverse action is that when holes and electrons recombine, energy is released. In germanium and silicon, this energy is released as heat into the crystal structure. However, in materials such as gallium arsenide and gallium phosphide this energy is released as visible light, and different semiconductor compounds release light of different wavelengths (colours). For example, infrared LEDs are made from silicon.

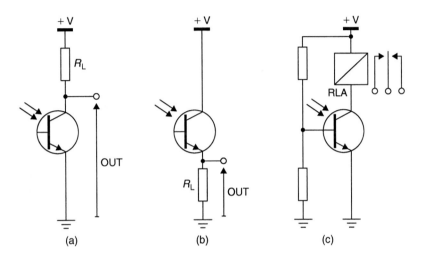

Figure 28.14   Phototransistor use: (a) collector load, (b) emitter load, and (c) driving a relay (back-emf diode omitted)

The basic structure of the diode is shown in Figure 28.15(a). The plastic moulding not only holds the component parts together, but also acts as a light-pipe so that most of the light generated is radiated from the domed region.

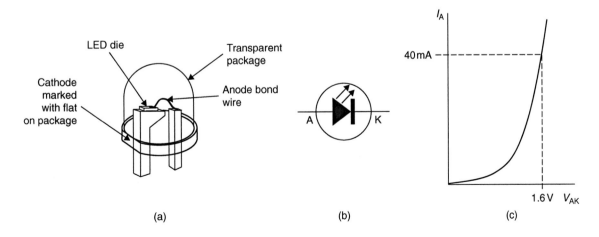

Figure 28.15   (a) Typical packaged LED, (b) symbol, and (c) forward characteristic

A characteristic typical of an LED is shown in Figure 28.15(c). When forward biased beyond about 1.6 V, the current rises rapidly and light is released. The current should be limited to a value less than about 40 mA by the use of a series resistor. In addition, the LED must not be reverse biased to more than a fraction of a volt, and circuits are often designed to avoid any possibility of reverse bias.

Opto devices

The optocoupler consists of an LED and phototransistor pair whose symbol is illustrated in Figure 28.16. The input signal modulates the diode current and hence the intensity of its (usually infrared) light output. This variation in light produces a variation in collector current to provide an output signal. Since the light beam has no electrical impedance, there are no matching problems between input and output circuits. The electrical isolation is very high, and an optocoupler can withstand test voltages as high as 4 kV between input and output terminals. Alternatively, an LED can be used to launch energy into a glass optical fibre cable to transmit the signal over very much greater distances. A variation of the optocoupler is the slotted optoswitch (Figure 28.16c). The light path between the LED and the photodiode can be interrupted by a shutter, generating a pulsating signal as the shutter breaks the beam of infrared light, which could be used to drive a counter-circuit, for example.

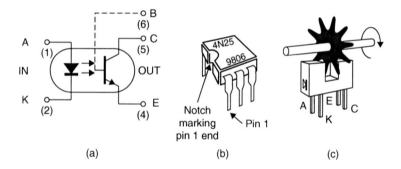

(a)   (b)   (c)

Figure 28.16   Optocoupler: (a) symbol, (b) typical six-pin dual in line (DIL) package, and (c) typical slotted optoswitch package

Photodiodes and transistors have similar fault tolerance to conventional semiconductors. Excessive voltage, current or temperature are the common causes of failure. Faulty devices can be located using the same techniques as applied to other semiconductors.

Laser diodes are used in many applications, from CD and DVD players to laser pointers, laser spirit levels and laser printers. The typical construction of a low-power laser diode of the type used in many of these applications is shown in Figure 28.17(a).

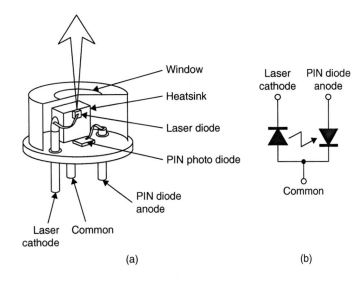

Figure 28.17 (a) Typical laser diode construction, and (b) schematic symbol

Warning: laser radiation can permanently damage your eyesight. Always observe the precautions on labels and in manufacturers' handbooks: never attempt to look directly at any sort of laser that is potentially turned on. Remember that although infrared light is invisible it can still damage your eyes. The symbol warning of laser radiation is shown in Figure 28.18. You will find this on CD players, laser printers and many other devices that use lasers.

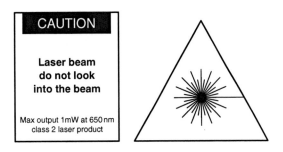

Figure 28.18 Laser radiation warning symbol

Typical laser diode applications that require a continuous output use feedback from the photodiode mounted in the same package (Figure 28.17) to close the loop of an automatic power control (APC) drive circuit. When the laser diode emits light, the photodiode current is proportional to light output power. This current is then used to control the drive to the laser diode, and this enables a constant output power to be maintained. Without such a power control circuit laser diodes can be vulnerable to thermal runaway and self-destruction. In pulsed applications the photodiode power control circuit is not always required, but is often necessary to reduce the effects of temperature, etc., on the output level.

Photovoltaic devices

In the classic type of photovoltaic cell, as used in camera light meters and similar devices, a layer of selenium is deposited on an iron or aluminium backing plate which forms the positive pole. A transparent layer of gold is evaporated onto this to form the negative pole. A metallic contact ring completes the circuit.

Light shines through the gold film into the layer of selenium. This releases electrons that form an electric field within the selenium, making the gold layer the negative pole of the cell. The whole cell is enclosed in a plastic housing with a transparent window for protection. The maximum current in bright light depends on the particular cell, but short-circuit currents in excess of 1 mA can be obtained. Such cells are suitable to drive a portable photographic light meter.

The majority of solar cells in use are silicon photovoltaic cells. These cells are typically manufactured by sawing lightly doped P-type silicon ingots into thin slices. The front surface of the slice has N-type material diffused into it to form a vertical P-N junction. The back of the slice is metallized usually with aluminium, and a grid or trellis structure of aluminium connections is deposited onto the front N-type surface (Figure 28.19a). Connections to these metal areas provide a way of collecting current from the cell. A cell of about 4 cm^2 is capable of providing 0.6 V on open circuit, with a short-circuit current of up to 100 mA in bright light. The cells can be connected in a series–parallel configuration to provide a greater output power level. The cost of manufacturing such cells has fallen steadily, and they are used extensively to provide power for applications as diverse as garden lighting and remote weather monitoring stations.

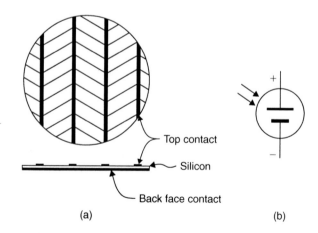

(a)   (b)

Figure 28.19   (a) Silicon solar cell, and (b) schematic symbol

Temperature sensors

Semiconductor diodes can be used as temperature sensors, as mentioned previously; however, they are typically only useful at low currents and over a relatively narrow temperature range, between about –60 and +180°C. Other temperature sensors can carry much higher currents: thermistors or thermocouples can operate at temperatures up to about +1300°C.

Thermistors

The thermistors may be in rod, bead, washer or disc form. Thermistors are made from carefully controlled mixtures of certain metallic oxides. These are sintered at very high temperatures to produce a ceramic finish. Thermistors have a large temperature coefficient of resistance, and it is this that makes them useful as a temperature sensor.

Negative temperature coefficient (NTC) thermistors have resistances that fall with a rise in temperature, and are commonly made from mixtures of the oxides of manganese, cobalt, copper and nickel. Positive temperature coefficient (PTC) components, whose resistance increases with a rise in temperature, can be made from barium titanate with carefully controlled amounts of lead or strontium.

Thermistors are used extensively for temperature measurement and control up to about 400°C. A typical heater controller (thermostat) circuit using an NTC thermistor is shown in Figure 28.20, in which the comparator circuit switches at some temperature set by the variable resistor VR1. Th1 is in good thermal contact with the object being heated. When the circuit is turned on (cold) the thermistor resistance is very high, so the voltage at the comparator inverting input is lower than the non-inverting input and the transistor is turned on, pulling in the relay contact and operating the heater. As the temperature rises the thermistors resistance falls, and a point is reached where the input to the comparator inverting input exceeds the non-inverting input voltage. The comparator turns off the transistor and heater. R5 provides hysteresis, feeding back a signal from the output to change the threshold

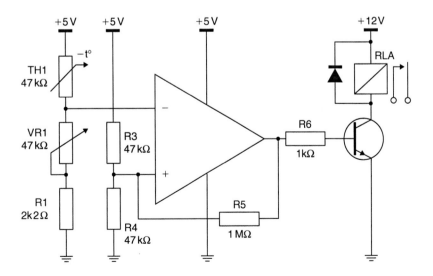

Figure 28.20 Typical temperature measuring circuit using thermistors

of the switching point. This prevents the circuit from oscillating near the set point. Look at the circuit and calculate the voltage at the non-inverting input of the comparator for both the heater on and off case.

Another application of thermistors is to provide temperature compensation for the change in the winding resistance of alternators and other generators, which affects their performance when the operating temperature rises. Similarly, NTC thermistors are used as in rush suppressors, to protect the cathode heater of a cathode-ray tube (CRT) at switch-on.

Thermoelectric transducers   When two dissimilar metals are in contact with each other a contact potential is developed between them. This is known as Seebeck or thermoelectric effect. The voltage, which rises with temperature, is almost linear over several hundred degrees. If two junctions are formed as shown in Figure 28.21(a), making a **thermocouple**, a current will flow around the circuit provided that each junction is at a different temperature. The circuit can be modified to include a meter which measures the difference in potential between the junctions. Accurate measurements can be made by holding one junction at a known temperature, usually the melting point of ice (hence called the cold junction), while the other becomes the temperature sensor. The two metals are chosen to maximize the contact potential for a particular temperature range. The cold junction in most applications is simulated by a semiconductor circuit called a cold junction compensator, rather than using a bucket of melting ice.

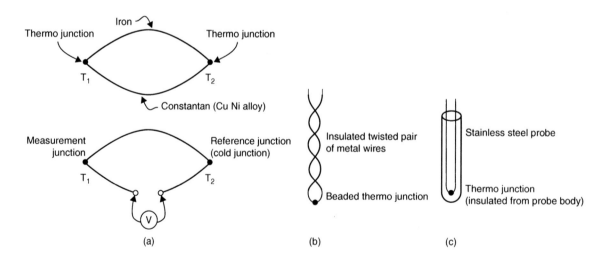

Figure 28.21   (a) Principle of a thermocouple junction, (b) construction of beaded junction sensor, and (c) construction of rod probe

The metals used include:

- iron and copper/nickel alloy
- copper and copper/nickel alloy

- nickel/chromium alloy and copper/nickel alloy
- nickel/chromium alloy and nickel/aluminium/manganese alloy.

Thermocouples allow for a range of instruments to measure temperatures from –85°C to 2000°C. For most temperature measurements, the hot junction is formed into a probe, either a low-cost bead on the end of a twisted pair of insulated wires (Figure 28.21b) or with a metal sheath (Figure 28.21c) to protect the junction from environmental hazards, and electrically isolated from it with magnesium oxide. This material has good thermal conductivity with high electrical resistance (up to 1000 M).

As noted earlier in this chapter, a temperature sensor can be based on the behaviour of a silicon diode. When a silicon diode is forward biased to operate at a constant current, its junction temperature changes linearly at a rate of –2 mV/°C. Therefore, this device can be used as a temperature-to-voltage transducer, where the output voltage can be used to indicate temperature.

Magnetic field sensors; the Hall effect

When a moving charge is exposed to a magnetic field, the path of the charge is deflected by interaction with the magnetic field. Current flow is due to a net movement of charge carriers, and the current carriers flowing within the conductor are deflected by magnetic field. Therefore, a voltage develops across the width of the conductor, and as a result a mechanical reaction occurs which tends to cause the conductor to move.

In normal conductors this voltage difference effect is of no practical importance, but when a magnetic field is applied to semiconductor materials, the amount of generated voltage becomes significant. It is put to practical use in a group of semiconductors called **Hall-effect devices**, illustrated in schematic form in Figure 28.22.

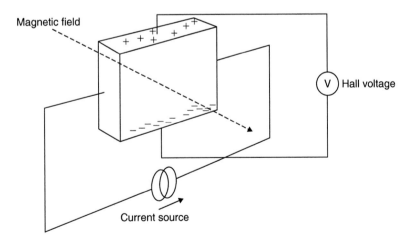

Figure 28.22 Schematic diagram of the Hall effect

A Hall-effect device consists of a very small (about $2 \times 2 \times 0.5$ mm) slab of gallium arsenide, indium antimonide or silicon onto which two pairs of electrodes are evaporated (Figure 28.22). A control current I and a magnetic field B are applied as shown, and the Hall voltage V_H is generated in a direction at right angles to both of them. The value of V_H is proportional to both the current I and the strength of the magnetic field B, depending on the dimensions and nature of the semiconductor material.

With a constant current passing, the Hall voltage is directly proportional to the strength of the magnetic field. The first and most obvious application for the device is therefore the measurement of the field strengths of both permanent magnets and electromagnets. It is also used in conjunction with switching transistors in brushless d.c. electric motors; by eliminating the mechanical brush gear that is so prone to wear, the reliability is improved. The device can also be used in contactless switches, where the movement of a permanent magnet sets up the Hall voltage, which in turn triggers a semiconductor switch.

When the application of the magnetic field in the Hall-effect device deflects the current carriers to one side of the conductor, they are caused to flow through a smaller cross-sectional area. This increases the effective resistance of the device, and so reduces current flow. This type of application is the **magnetic-dependent resistor** (**MDR**). Over a normal range of magnetic fields strengths, device resistance can in this way be made to vary by a factor of about five. One particularly useful application of an MDR is in the 'clamp-on' type of current meter used to measure current flow in power supplies drawing power directly from the electrical mains.

Solid-state relays

Solid-state relays are typically used in mains power control applications where low-voltage devices such as microprocessors are required to operate high-voltage, high-current equipment. Typically having an optically isolated input stage (Figure 28.23) and triac or dual thyristor zero-crossing switched output stage makes solid-state relays much more reliable than their mechanical equivalents. Output current ratings of 10–125 A at 230 V r.m.s. are commonly available.

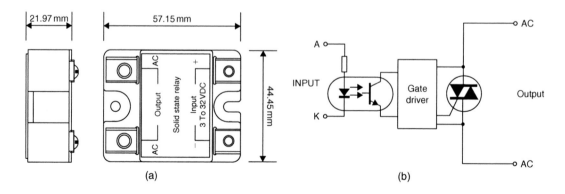

Figure 28.23   (a) Typical solid-state relay package, and (b) simplified circuit

Electrostatic discharges

Lightning and ionospheric storms are the chief sources of high levels of electrostatic discharge (ESD), but artificial electrostatic charges are rather more subtle. These can be generated by the rubbing together of two dissimilar materials. Actions such as walking across a carpet or even the nature of the clothing worn can give rise to triboelectricity, as this phenomenon is known.

The discharge of static electricity can easily damage MOSFETs, complementary metal-oxide semiconductors (CMOS) and group III/V compound semiconductor (GaAs, etc.) devices. A discharge from a potential difference as low as 50 V can cause component degradation, and since the effect is cumulative, repeated small discharges can lead to ultimate failure. It is therefore important that any sensitive equipment should be serviced in a workshop where static electricity can be controlled. Bipolar transistors are rather more robust in this respect.

Note: this is usually a problem only when individual devices or modules such as computer memory are being handled. Devices that are wired into a circuit usually have comparatively low resistance between electrodes, so damage is unlikely. The problem is much more acute for manufacturers than for service personnel, particularly since much of the servicing of circuits that contain components liable to ESD damage will be done by replacing modules rather than individual devices.

For working with such components the basic workstation should provide for operators, workbench, floor mat, test equipment and the device under service to be at the same electrical potential. Operators should be connected to the workbench via a wrist strap. To avoid static electricity the operator's clothing also needs to be considered. The wearing of wool and artificial fibres such as nylon creates considerable static. One very useful garment is a smock made of polyester fabric, interwoven with conductive carbon fibres. This has through-the-cuff earthing. The use of compressed air to clean down boards can generate static and this can be avoided by the use of an ionized air blast.

Static-sensitive components should be stored in conductive film or trays until required and then handled with short-circuited leads until finally in circuit. Soldering iron bit potentials can also be troublesome unless adequately earthed. The workbench surface should be clean, hard, durable and capable of dissipating any static charge quickly. These static-free properties should not change with handling, cleaning/rubbing or ambient humidity.

The use of an ionized air ventilation scheme can be an advantage. Large quantities of negatively and positively charged air molecules can quickly neutralize unwanted static charges over quite a significant area. Capacitors, including the capacitance of CRTs, should not be discharged too rapidly, to avoid damage through excessive current flow. Normal power supply capacitors can be safely discharged through a 250 V rated 0.5 W 1 K resistor, but for the extra high tension (EHT) voltages found on the CRT, it is safer to use a 100 K resistor.

Multiple-choice revision questions

28.1 A transistor has a stated value of h_{fe} equal to 150. What base current (in microamps) would you expect to produce a collector current of 60 mA?
   (a) 150 μA
   (b) 400 μA
   (c) 60 mA
   (d) 9 A.

28.2 A circuit operating from a 5 V supply uses a silicon bipolar transistor in common emitter configuration to turn on a relay. What is the base emitter voltage when the relay is on?
   (a) 0 V
   (b) 0.2 V
   (c) 0.6 V
   (d) 5 V.

28.3 To turn off a P-channel enhancement MOSFET, the gate voltage must be?
   (a) equal to the drain voltage
   (b) higher than the source voltage
   (c) lower than the source voltage
   (d) 0 V.

28.4 What is the gate–cathode voltage is required to turn on a thyristor (SCR)?
   (a) 0.6 V
   (b) 1 V
   (c) 6 V
   (d) 230 V.

28.5 A light-dependent resistor is use to turn on a night light at dusk. What resistance would you expect it to have in bright sunlight?
   (a) >10 MΩ
   (b) <1 Ω
   (c) <100 kΩ
   (d) 50 Ω.

28.6 A red LED is to be operated from a 13.8 ±2 V supply. If it has a VF of 1.6 V and must draw <20 mA from the supply, what resistor would you use to limit the current?
   (a) 220 Ω
   (b) 390 Ω
   (c) 820 Ω
   (d) 10 kΩ.

Unit 2

Outcomes

1. Demonstrate an understanding of analogue and digital meters
2. Demonstrate an understanding of oscilloscopes and their uses
3. Demonstrate an understanding of test and measurement instruments for components and signals
4. Demonstrate an understanding of the role of the PC in test and measurements
5. Demonstrate an understanding of equipment reliability and surface mount repair
6. Demonstrate an understanding of safety testing and electromagnetic compatibility.

29   Meters and data

Since the servicing of all electronic circuits calls for the use of measuring instruments, it is necessary to be able to make effective use of these and of the readings obtained from them. The two most important types of circuit measurements are d.c. voltage and current readings, using a multirange meter (multimeter), and signal waveform measurements that are best made using a oscilloscope. One preliminary point of importance is that, whatever the type of measurement made in an electrical or electronic unit, the very act of connecting the instrument into the circuit will have some influence on it and so affect the reading obtained.

Apart from the ease of reading an instrument and its ability to produce consistent readings, the most important features of any test instrument are its accuracy, resolution and linearity.

The **accuracy** of any meter is quoted in percentage of error relative to some particular standard value. If a meter has a declared accuracy of 5% (say 5 mV in 100 mV), there is little point in trying to interpret a reading to an accuracy of 1 mV. As a useful guide, read only to the nearest half division on the indicated scale. A high-grade analogue meter usually has a mirror-backed scale plate so that the pointer can be accurately read by avoiding parallax errors. The operator's head is moved from side to side so the reflection just disappears behind the pointer (Figure 29.1b). A digital multimeter often has a better accuracy than an analogue one that uses a moving coil unit. For general-purpose instruments, an analogue multimeter may have a declared accuracy of 0.5%, while the corresponding digital instrument may produce a figure of 0.025%. However, the least significant digit often varies while taking a reading, so that the accuracy would typically be declared as $0.x\% \pm 2$ counts.

Resolution refers to the smallest distinction between readings that can be obtained. For the analogue instrument this is typically the nearest half division mentioned above. For the digital device it is the value represented by the least significant digit.

Linearity represents a scaling that varies by equal deflections per unit of measured quantity. For an analogue instrument with poor linearity this is usually obvious by the cramping of scale readings towards one end. The linearity of a digital multimeter is not similarly obvious from the reading of the display, but can still exist as an unwanted feature of the measurement system.

Periodically, the accuracy of all instruments should be checked against some standard voltage cell or an instrument that has been similarly calibrated and can be used as a transfer standard. Since this property is often temperature dependent it is usual to make any readjustment at 20°C. The calibration against a transfer standard meter can easily be affected by connecting the meters to be compared either in parallel (voltage) or in series (current) when coupled to the same source.

Unlike multirange instruments used for service purposes, **panel meters** Figure 29.1(c) are often built into control systems monitors. While these meters may well be scaled in volts, amps or ohms, they may equally well be scaled in degrees Celsius, kilograms or any other parameter such as litres of liquid flow per second. They may also contain a limited number of shunts and multipliers so that the single meter can be switched to indicate a small number of different parameters. Otherwise, analogue panel meters are simply basic instruments used for a very specific purpose. In most cases modern panel meters are microprocessor-based systems which can include features such as event recording and replay, peak hold and alarm outputs or even network connections.

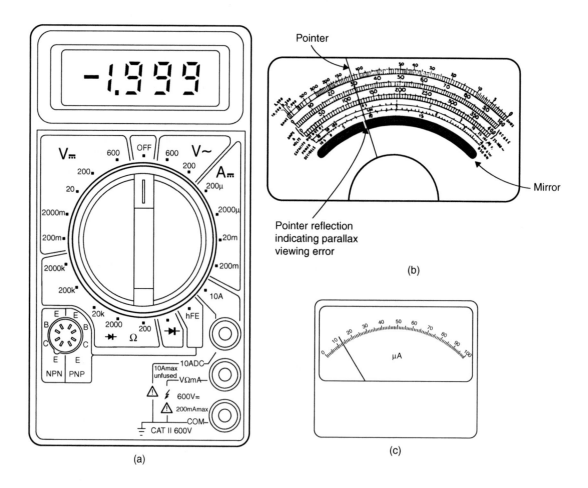

Figure 29.1   (a) Typical low-cost digital multimeter, (b) mirrored scale analogue multimeter scale, and (c) 100 µA moving coil panel meter

Multimeters are instruments capable of measuring several ranges and parameters, often d.c. voltage and current, resistance and a.c. voltage. Some multimeters include a.c. current scales, usually more expensive types, because such readings have an acceptable accuracy only if the meter includes an expensive current transformer, or 'true r.m.s.' analogue or digital signal processing circuitry.

Both analogue and digital multimeters (Figure 29.1a) are in general use. Analogue multimeters use a moving-coil movement which, because of its resistance, draws some current from the circuit under test. Digital multimeters, in contrast, contain no moving parts; the reading appears as a figure displayed on a readout similar to that of a calculator. A separate power supply, usually a battery, is used, so that practically no current is drawn from the circuit under test. Either type of meter will have a measurable input impedance whose value will affect the readings. Analogue meters referred to as field-effect transistor input volt ohm amp (FET VOM) combine the input amplifier circuitry of the digital multimeter with a precision moving coil meter. These instruments are usually found in specialist laboratory applications.

Practical 29.1

Connect the circuits shown in Figure 29.2 and use first an analogue and then a digital multimeter, each set for a 10 V range, to measure the voltage *V*. Note the results, and check by calculation that the value of *V* ought to be 4.5 V when the meter is not connected. Are the meter readings significantly different for (a) the low-resistance circuit, and (b) the high-resistance circuit? Analogue meters are described as having a input resistance of so many ohms per volt; compare this rating with the circuit measurements you have made.

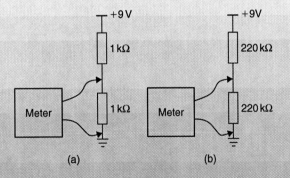

Figure 29.2 Circuit for comparing meters: (a) low resistance, and (b) high resistance

As the results of the practical show, the lower resistance of the analogue meter can cause the readings taken to be unreliable. These voltage readings will always be in doubt when the resistance of the meter itself is not high compared with the resistance across which the meter is connected. For reliable readings, the resistance of the multimeter must be high compared with the value of resistance at the point where the measurement is taken. At least 10 times higher is the minimum, and even at this level, you can expect to find some difference between the true voltage (when no meter is present) and the measured value. The digital meter, with its high resistance, is definitely superior to an analogue type in this situation.

Range extension using multipliers and shunts

The basic meter resistance is that of the moving coil, which is both temperature dependent and subject to significant manufacturing tolerance. To reduce the effects of these variations on accuracy the meter is usually connected in series with a much larger resistance to swamp the variations in coil resistance.

If a basic meter with 50μA full-scale deflection (FSD) and 1 kΩ resistance is to be used to measure voltage, then the largest voltage that can be measured is 1 kΩ × 50μA = 50mV. In order to measure a higher voltage, say 10V, a series multiplier resistor needs to be connected in series with the meter (Figure 29.3a). The multiplier needs to drop 10V–50mV when 50μA flows through it, so its value is 199kΩ. This gives a total resistance of 200kΩ for the 10V meter. The meter is then said to have an input resistance of 20kΩ/V.

Extending the current range of a meter can be done in a similar way, using a parallel shunt resistor Figure 29.3(b). Taking the same 50μA meter movement with 1 kΩ resistance as the previous example, we can use it to measure a current of 2A by selecting a shunt which drops 50mV when 2A flows through it and the 1 kΩ meter resistance in parallel, that is 50mV/2A = 25 mΩ. The total resistance is thus 25 mΩ, composed of 1 kΩ in parallel with a 25.000625 mΩ shunt, which would be a short thick piece of copper wire. If you look inside a multimeter this is exactly what you will find between the A input and the common terminal. Note in this case that the difference in meter and shunt resistance is so great that the effect of the meter resistance is negligible and a 25 mΩ shunt would be used.

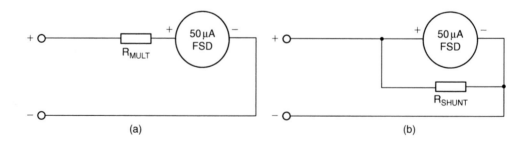

(a)   (b)

Figure 29.3   (a) Using series 'range multiplier' resistor to extend the voltage range of a moving coil meter, and (b) current shunt to extend the current range

Resistance measurement

The basic analogue meter can be adapted to add a resistance range to the multimeter circuit. The battery, shown in Figure 29.4 as 1.5V, is used to drive a series current through the meter and the resistance under test. The variable resistor R1 is used to set the meter to FSD with a short-circuit across the test terminals which represents the point of 0 ohms. The open-circuit condition should produce a meter deflection of zero and this represents infinite resistance (∞). The meter now measures in the reverse direction.

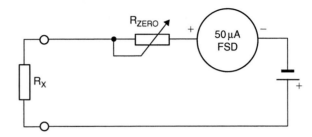

Figure 29.4   Adding a resistance range to a meter

The range of such an instrument can be modified simply by increasing the battery voltage to, say, 3 V. Meters of this type normally have a positive potential on the common terminal black test lead and vice versa. It is thus important to recognize this point, particularly when measuring non-linear devices such as diodes and transistors. A similar system can be used to provide a resistance range on digital multimeters, with the advantage that much higher resistance values can be measured.

The simple circuit known as the **Wheatstone bridge** can provide accurate measurements over a very wide range of values. As shown in Figure 29.5, it consists simply of three resistors, one of which is variable, a centre zero moving coil d.c. meter and terminals across which the unknown resistor is connected. The bridge is energized by a low d.c. voltage (usually a 1.5 V cell). The resistors R1 and R2 are referred to as the **ratio arms**, which can be switched in or out of circuit as necessary to provide resistance ratios of R1:R2 of, typically, 10:1, 100:1 and 1:100, 1:10. The unknown component is connected to provide Rx and the bridge circuit is balanced when there is zero current through the meter (at centre scale). The balance condition occurs when R1Rx = R2R3, or when Rx = R2R3/R1.

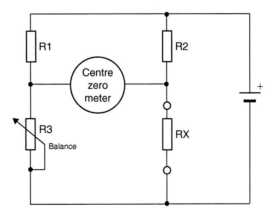

Figure 29.5   Basic Wheatstone bridge circuit

The ratio arm values are known and the value of R3 is found from the scale surrounding the control knob. This allows a simple calculation to be made which provides the value of the unknown to a very high degree of accuracy. The bridge type of circuit is used extensively for measurement and is also the basis of the full-wave rectifier circuit.

By connecting a d.c. moving coil meter to a bridge rectifier circuit, a.c. voltages are converted into d.c. (Figure 29.6). The d.c. meter itself can be fitted with shunt and multipliers so that its basic range can be extended. When used in this way, the meter reads the average value of the sine wave input, equal to 63.7% of the peak value. By modifying the scaling of the meter, this can be redrawn to indicate the r.m.s. value instead. Since the r.m.s. value is 70.7% of the peak a.c. value, this scaling or form factor for a sine wave is equal to 0.707/0.637 = 1.11. Different form factors are needed for other waveforms. It should be noted that the bridge rectifier introduces a voltage drop of 1.4 V or so into the circuit, so this circuit is inaccurate for a.c. voltages much below 20 V.

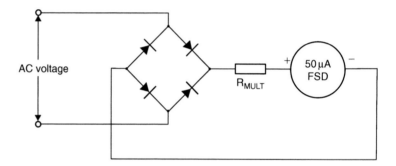

Figure 29.6   Measuring a.c. voltage with a bridge rectifier

When shunts similar to those used for modifying the ranges of a d.c. meter are used for a.c. measurements, their impedance, which varies with changes in the magnitude of the current, would require the use of separate scales for each range. This is the chief reason why a.c. shunts are only found in very low-cost instruments. To overcome this problem, a **current transformer** (Figure 29.7b) with a tapped secondary winding is used instead. This has a low-impedance primary winding and a high-impedance secondary, so that the secondary voltage is proportional to the primary current.

For example, If the basic 50 μA FSD meter were to be used to measure 5 mA a.c., then the transformer would need to have a 1:100 turns ratio. In this case, the conversion between average and r.m.s. scaling can be accomplished within the current transformer by multiplying the turns ratio by the standard form factor of 1.11. Thus, the required turns ratio becomes 1/100 × 1.11 = 1:90. The a.c. output voltage is finally rectified using a standard bridge circuit as shown in Figure 29.7.

In theory, the shunts and multipliers used to extend the ranges of basic analogue instruments could be chosen from separate, high-stability, high-accuracy standard components. However, if these separate resistors are switch selected while the meter is left connected to a circuit under test, there is a very high

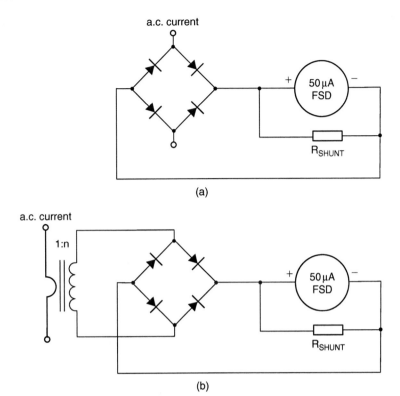

Figure 29.7   Measuring a.c. current: (a) low current, and (b) high current measurements using a current transformer

chance that the full test voltage or current can be momentarily applied across the meter, with disastrous results. To avoid this situation, shunts and multipliers are generally wound in a series configuration and tapped at the appropriate points. By using such **universal shunts** or multipliers, the problem mentioned above can be nearly completely avoided.

Digital multimeters (DMMs) generally have much less effect on the circuit being measured. Most of them have a constant input resistance of 10M or more, and few circuits will be greatly affected by having such a meter connected into them. The operating principle is that the input voltage is applied to a high-stability, high-resistance potential divider which is usually connected to an operational amplifier (opamp) to provide further isolation. The output of the opamp then provides the input to an analogue-to-digital converter stage (Figure 29.8). Typically, a microprocessor circuit drives the display, provides functions such as peak hold, and can also output data to an RS232 serial link or other data communication device. The range switch selects the part of the potential divider to be used, and the position of the decimal point on the display.

A point to note is that although a digital meter may indicate a voltage reading to several places of decimals, this is not necessarily more precise than the reading on an analogue meter.

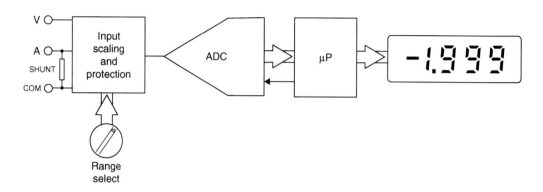

Figure 29.8   Simplified block diagram of a digital multimeter

DMM devices are often specified by the number of digits in the display. However, there is an extension to this technique to allow for over-range readings. An extra leading digit is either displayed as a 1 or it is not displayed. Such a DMM would be described as having a 3½ digit display. A further addition provides for a minus sign (–) to be displayed when reverse polarity is encountered. This adds an extra ¼ digit. Therefore, a basic three-digit display with these features would be described as a 3¾ digit meter.

Meter features

Frequency response becomes important for a.c. measurements. While most multimeters are built to monitor power frequencies, they can provide a reasonably accurate assessment of voltages or current, up to about 20 kHz for an analogue instrument and perhaps 100 kHz for a digital multimeter.

Autoranging allows a multimeter to select automatically the correct range of values, when the specific function has been preselected by the user. The feature cannot easily be built into an analogue instrument, but can readily be included in the facilities of a DMM, particularly those instruments that are software controlled. Basically, if the reading on a particular selected scale has been overdriven, the meter automatically shifts the scaling up to the next highest level. However, overload can still occur if the meter tries to exceed the maximum permitted level.

Capture of readings applies chiefly to the DMM and represents the time taken for a reading to reach equilibrium. For the analogue instrument the pointer settles in a time controlled by the damping created by the back-electromotive force (emf) due to the pointer movement. In the DMM, this feature can be affected by the dither in the least significant digit (±1 or 2 counts).

Peak readings: analogue meters measure the average value of the input quantity, which in the case of d.c. is also the peak value. However, for a.c. quantities, the scaling is modified by the form or scaling factor. Digital meters can be designed to measure true r.m.s. values for any waveform.

Clamp-on current measurements: when attempting to measure the current flow through, say, a large power cable, it is not convenient to break the cable to insert a meter. For these applications there is a device which is effectively the secondary of a current transformer wound on a split core that can be opened to encompass the cable and then read off the current flow directly. Such an instrument can include within the split core clamp a Hall effect device that responds to the magnetic effect of d.c. current flow so that the meter interprets the strength of this field in terms of the d.c. current that created it, and this allows both a.c. and d.c. current measurements to be made. It is important to ensure that only one wire, e.g. the supply (or live for a.c. mains) wire or the return (or neutral) wire, but not both, passes through the clamp, otherwise the currents in the two wires will cancel each other out.

Transistor and diode tests: digital multimeters often have facilities for testing transistor h_{fe} and diode forward voltage drop. These can be useful for matching devices, but are not usually very accurate. The transistor h_{fe} measurement usually operates by applying a small fixed current, e.g. 100 μA, to the transistor base emitter junction and measuring the current flowing in the collector circuit. Similarly, diode forward drop operates by trying to force 5 mA through the diode and measuring the voltage required to do so.

Overload protection: there is a number of steps that the user can take to avoid damaging a meter through overloading:

- Take note of the manufacturer's stated safe working levels for both a.c. and d.c.
- Avoid measuring voltages or currents that exceed the maker's recommendations on any particular range.
- Avoid trying to measure volts or current with the meter switched to the resistance range.
- Observe the stated safe environmental working temperature.

Apart from the obvious use of a fuse in the test lead circuit, particularly for current ranges, back-to-back silicon diodes (**antiparallel diodes**) are often included in parallel with the meter movement. These develop a low resistance on overload and thus place a short-circuit across the meter movement. Analogue meters may have a ballistic mechanical system that physically trips a pair of contacts to remove it from the circuit if the pointer approaches the end stop too quickly. Moving coil meters should always be stored or transported with a short-circuit across their terminals. This damps the movement due to the back-emf created by physical movement. Electronic driven digital or analogue meters can be protected by using the crowbar effect of a pair of thyristors wired across the input.

Because of the flexibility of the DMM, its operation can easily be extended to provide measurements of capacitance, frequency, sound level, temperature, humidity and other parameters, as well as allowing for simple diode and transistor testing.

Practical 29.2

Connect two meters, one digital and one analogue, to the centre-tap of the potentiometer, illustrated left. Switch the meters to their 10 V ranges and switch on the 9 V supply. Observe the readings as the

(Continued)

Practical 29.2 (Continued)

potentiometer shaft is rotated to and fro. Which meter more closely follows the changes in the output? Check also how quickly you can read each meter when the voltage is steady. Which meter is easier to read?

Making measurements

- Start with the meter switched to its highest voltage range.
- Connect the meter with the circuits switched off.
- For voltage measurements, always use the highest range that gives a readable output.
- For current measurements, always try the highest current range first.
- Never leave the meter switched to any current range when you have temporarily stopped using it.
- Make sure you know which scale on the dial to read before you try to take a reading.
- Never leave a meter switched to ohms range when taking voltage or current readings.

In most circuits, the use of a multimeter for current readings has much less effect on the circuit than when it is used for voltage readings. To make a current reading, however, the circuit has to be broken, and this is seldom easy on modern circuit boards.

A few audio and television circuits make provision for checking currents by having a low resistance permanently connected to the current path. By measuring the voltage across this resistor, the current can be calculated using Ohm's law, $I = V/R$. Because the resistance so placed in the circuit is very small, the effect of connecting the meter into the circuit is negligible.

Voltage readings are used to check the d.c. conditions in a circuit. These readings are usually made when no signal is present. Readings that are taken when a pulse signal is present will **always** give misleading results because of the effect of the pulse signal itself on the meter. For this reason, voltage readings shown on a circuit diagram are usually specified as 'no-signal' voltage levels, or are shown only at points in the circuit where the signal is decoupled so that only d.c. is present.

When voltage readings are shown on a circuit diagram, they are always average readings which may vary from one example of a circuit to another because of component tolerances. In addition, the actual readings taken in a circuit may be different from these values, either because of tolerances or because of the use of a meter with a different input resistance, even if the circuit is working quite normally. Some experience is needed to decide whether a voltage reading which is higher or lower than the stated average value represents a fault, or whether it is simply due to tolerances or meter resistance. In general, a voltage varying from the norm by up to 10% is usually due to tolerances, but a voltage variation that shows a transistor to be nearly bottomed or cut-off in a linear stage always betrays a fault condition.

Root mean square values

When d.c. flows through a circuit it dissipates heat due to the power being generated, I^2R W. Similarly, when a.c. flows through a circuit power is again dissipated, but how is this to be compared and measured? For a sine wave, the average voltage or current flowing will be zero, because each positive excursion is matched by an equal but opposite polarity negative swing. But we know that even a resistor passing a.c. gets hot, so how can we measure this effect?

The root mean square (r.m.s.) value is described as that value of a sine wave that produces the same power or heating effect as an equivalent d.c. value. For a sine wave, this can be calculated first by squaring the waveform and then taking the average value of this, either V^2 or I^2. Now, by taking the square root of this value, either $\sqrt{V^2}$ or $\sqrt{I^2}$, we return to a particular value of V or I, which is termed the r.m.s. The process is shown in Figure 29.9.

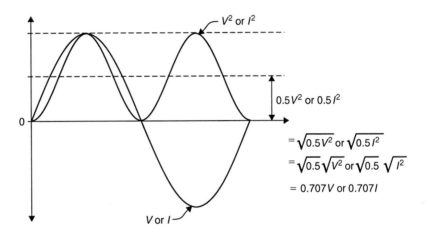

Figure 29.9 Derivation of r.m.s. value for a sine wave

Since the power dissipated or the heating effect is only truly applicable to the sine wave a.c. drive, any other wave shape would produce different results. For example, it can easily be shown using the technique explained above that the average, peak and r.m.s. values for a square wave are all identical. A method that can be adopted for calibrating a non-sinusoidal alternating waveform relies on the basic definition. By using the power of an unknown wave shape to drive an incandescent lamp and a controllable true sinusoid (or d.c.) to power an identical lamp, increase the drive to and monitor the sine wave drive r.m.s. level. When the two lamps glow with equal brightness as measured by a light meter or temperature sensor, the r.m.s. value of the arbitrary wave shape is equal to that of the sinusoid or d.c. drive power.

Networked instruments

The concept of automated test equipment (ATE) has been progressively introduced into the servicing environment, and this involves simple local area networks (LANs). Not only can the networked system now provide service information to many test and repair stations, such systems can also be used to monitor and log data from various test points for future analysis. Furthermore, the computer monitor can be converted into a range of test instruments, all under software control.

RS232 is probably the oldest interconnection standard for linking a computer to an item of ancillary equipment. During its more than 25-year existence, the standard has undergone more than six major changes and is still valid today. The original RS232 standard was designed to provide two transmit/receive channels for simultaneous full duplex communications using bipolar signals of $\pm 15\,V$ level. The positive and negative excursions represented logic 1 and 0, respectively. Using a multicore cable and 25-pin D-type connectors, a maximum signalling rate of 20 kbits/s could be achieved over a distance of 20 m. Later versions of RS232 (such as RS423) have used both 15-pin and nine-pin D-type connectors with multicore cable and bipolar signals at both ± 3 and $\pm 5\,V$ levels in a single simplex transmit/receive mode.

In all the RS232 versions, the system transmits in the bit serial mode in both directions and the system standard provides for data, timing and control signals. The system is designed to connect a **data terminal equipment (DTE)**, usually a computer, a **data communication equipment (DCE)** or a peripheral device. The **primary channel** is normally used for data, while the secondary path carries the control and timing signals. However, both channels may be configured for both half-duplex and full-duplex transmission modes.

Figure 29.10 shows the standard pinning arrangement for the RS232 DTE connector, and Figure 29.10(c) shows the full implementation of the pin allocation. Few applications now make use of all of these pins.

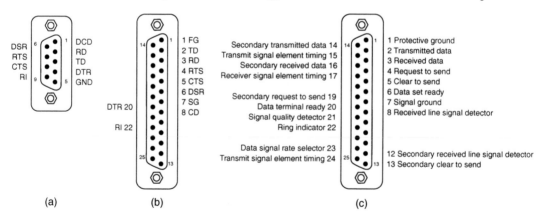

Figure 29.10   RS232 pin connections: (a) nine-pin DTE, (b) 25-pin DTE, and (c) full 25-pin DTE

For a modem cable using the full 25-pin implementation, the two connectors are configured pin for pin, but for the **null** or **non-modem** operation, pins 2 and 3 (RXD and TXD) and pins 5 and 20 (DTR and CTS) are cross-connected. This type of cable is used, for example, when the cable is connecting two computers together so that they can pass data in either direction. The changes for the nine-pin cable are pin to pin for a modem cable, and for the null-modem, pins 2 and 3 (RXD and TXD) and pins 4 and 8 (DTR and CTS) are cross-connected.

The **IEEE488** data cabling system is also known as **general-purpose interface bus (GPIB)** and **Hewlett–Packard instrumentation bus (HPIB)** and is a standard interface bus designed specifically for instrumentation purposes. It consists of eight data lines: three control lines and five lines for interface control. It is based on a standard male/female back-to-back 24-pin connector and cable system, which provides for 8-bit parallel data transfer rates of 1 Mbyte/s. The transmission system uses unipolar negative logic signalling (high = 0, low = 1). Fifteen devices with different data rates may be connected to the bus at any one time, but the total transmission line length is restricted to 20 m maximum.

In the latest version, IEEE488.2, the parameters have been upgraded. Three types of device may be connected and these are described as talkers, listeners or controllers. For example, a meter, keyboard or sensor may be described as a talker, a printer or a recorder as a listener and a computer as a talker–listener. By comparison, a DMM that needs to be programmed in situ and then be able to transmit results is also a talker–listener. Although the network may contain more than one computer, one is designated as a master or controller and allocates the bus to individual devices in turn.

Each device on the bus has its own unique address carried in a specific byte. The five least significant bits (LSBs) represent the address, bits 6 and 7 set the talk/listen function, with 01 = listen and 10 = talk. The most significant bit (MSB) can generally be ignored.

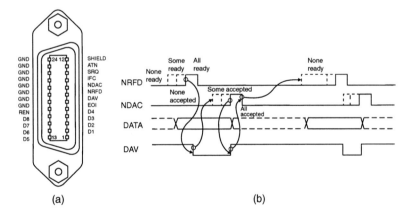

(a)   (b)

Figure 29.11 IEEE-488 connector: (a) pin out, and (b) timing of a talker–listener IEEE-488 exchange. Where more than one listener exists, DAV is not asserted until all listeners are ready.

- **ATN** (attention) is the main controller signal. When this is low, the controller is sending commands or addresses to the interface connected devices. When it is set high, it signifies to the talkers and listeners that the bus is ready to transfer data.

- **IFC** (interface clear): by driving this signal low for a short period ($100\,\mu s$) the controller resets the interfaces of all the devices on the bus.
- **REN** (remote enable): this line is held permanently low when the interface is working.
- **SRQ** (service request): devices use this line to request service from the controller. For example, an instrument can signal that it is ready to read after taking a measurement.
- **EOI** (end or identify): this line is driven low by an active talker while it is transmitting its last byte as a form of end of data string. It can also be used by the controller.
- **DAV** (data valid).
- **NDAC** (not data accepted).
- **NFRD** (not ready for data).

When ATN is low (1), all devices must listen for interface messages, then if a device is addressed, it must be ready for a data transfer as soon as ATN goes high (0). Every byte is transmitted under the control of three handshake lines, DAV, NRFD and NDAC.

An active talker with data to transmit monitors the NRFD line and if this is low (1), the listeners are not ready to receive data. When NFRD goes high (0) and the listeners maintain NDAC low (1), the data is not being accepted. When the listeners set NDAC high (0) and NFRD low (1), the data can be transferred.

At the end of each transmitted byte DAV goes low (1) to validate the data transfer.

Multiple-choice revision questions

29.1 Given an equal specification, what advantage makes digital meters easier to use than analogue meters?
(a) larger display
(b) no parallax errors
(c) does not load the circuit
(d) does not require a battery.

29.2 Using the 2 V range of a 20 000 Ω/V analogue meter, a voltage of 0.4 V is measured across a 100 kΩ resistor in circuit. What is the real resistance in circuit?
(a) 20 kΩ
(b) 29.6 kΩ
(c) 71.4 kΩ
(d) 100 kΩ.

29.3 Using a 1 mA FSD meter with an internal resistance of 1 kΩ, what shunt resistor would be required to measure 4 A d.c. at full scale?
(a) 4 Ω
(b) 1 Ω
(c) 0.25 Ω
(d) 0.025 Ω.

29.4 A current transformer is used to measure the current in a circuit. What turns ratio should be used to measure 2 A r.m.s. on a 1 mA FSD 1 kΩ meter?
(a) 1:1400
(b) 1:1800
(c) 1:2000
(d) 1:2200.

29.5 Which of the following features is not found on an analogue meter?
(a) auto ranging
(b) 10 MΩ input resistance
(c) true r.m.s. current measurement
(d) transistor hfe tester.

30   The analogue oscilloscope

The **cathode-ray oscilloscope** (**CRO**) (Figure 30.1) is a device that is capable of writing a two-dimensional (2D) and, in some cases, a three-dimensional (3D) graph on a glass screen. It does this under the influence of two (or three) input signals whose effect is to produce a recognizable pattern that describes certain features of the various signals. At this point you should revise these topics in Chapters 11 and 12 (Level-2 Book). Oscilloscopes are the mainstay of circuit testing because circuits are usually designed to operate on signals that change with time and very few, apart from power supplies, are designed not to change with time.

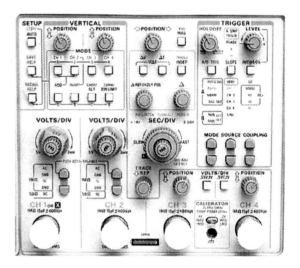

Figure 30.1   A typical analogue oscilloscope, the Tektronix 202445A

The older type of oscilloscope is purely analogue in nature and to overcome the transient nature of the display, the CRO can be fitted with a camera attachment to provide a permanent record. Each oscilloscope type is commonly available with a range of alternate cathode-ray tubes (CRTs), each with a different persistence of illumination and colour.

At the heart of the display system is the **sawtooth** waveform that is used to produce the horizontal beam deflection across the tube face. This causes the beam to traverse the screen relatively slowly during the forward writing period and then rapidly fly back to repeat the process in a continuous manner. While this is in progress, the signal to be examined, the work signal, is applied to the vertical deflection system so that a 2D pattern that describes its amplitude, periodic time and general shape can be displayed. Typical basic beam-deflection sensitivities, which are accurately known for each tube, vary between 0.02 and 0.05 cm/V.

A CRO with a single-beam CRT can function as a multitrace CRO in one of two modes, **chopped** or **alternate**. In the chopped mode, the beam is rapidly switched many times per trace cycle between two channels to provide a double-trace display. Switching high-frequency signals in this manner produces unacceptable gaps in the displayed waveform and thus this mode is limited to use at relatively low frequencies. In the alternate mode, one sweep is used alternately for each channel, but at the expense of reduced trace brightness. True **double-beam CRT**s use tubes with two identical electron guns mounted side by side within a common glass envelope. These can also be adapted as above to provide four-channel operation.

Using an analogue oscilloscope

The signal to be examined is input to the Y-amplifier section through a calibrated step attenuator which commonly provides an input impedance ranging from 1 to 10 M in parallel with a self-capacitance of 10–20 pF. A few specialized instruments use a 50 Ω input impedance. This stage typically has a rise time of a few tens of nanoseconds (rise time is the time taken for the signal to rise or fall between its 10% and 90% levels). The input bandwidth can easily exceed 100 MHz. Dual-channel CROs will be equipped with two such stages. The major part of the Y channel gain is achieved in the driver and output stages to the Y-plates.

The X-deflection system is driven from a sawtooth generator stage that provides a waveform with very low distortion. This signal is further amplified before being used to drive the X-deflection system. Again, for a dual or double beam instrument there will be two similar display driver stages, but generally driven from one (or possibly two) timebase circuits. The timebase sawtooth can be synchronized to either of the two input signals (Y) selected by a switch, as shown in Figure 30.2. Provision is also commonly available to compare the Y input with an external signal and therefore an additional switched input can be provided as shown. In addition, d.c. shift potentials are provided for both the X and Y channels to position the trace precisely within the graticule scale, which is commonly ruled in 1 cm squares.

At times, it is necessary to study just a short time-scale section of the Y signal and this can be done by blanking this input until such times as the important section becomes available. This can then be studied by illuminating the trace through a bright-up signal known as the **Z modulation**, so that only the wanted period is displayed. The effect is almost that of a 3D display.

The visual display is supported by an engraved and sometimes illuminated measuring scale called a **graticule**. This allows fairly accurate assessments of amplitude, phase relationship and time to be made. In addition, with experience it is possible to obtain a reasonable assessment of any distortion found in a waveform.

Normally, the Y input is coupled via a high value of capacitor to ensure that the frequency response can extend as low as 10 Hz. The CRO is

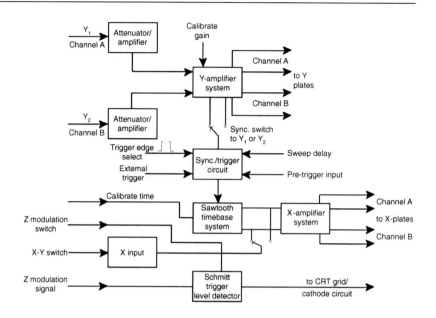

Figure 30.2　Block diagram of two-channel analogue oscilloscope

equipped with a switch labelled **a.c./d.c./ground** and in the d.c. position the coupling capacitor is short-circuited. This is provided so that the Y input signal that has a d.c. component can be displayed correctly. The ground position is useful because it sets the unmodulated horizontal trace at ground voltage level, which allows this axis to be accurately positioned on a particular graticule line referenced as zero volts.

Even with relatively low-cost oscilloscopes several variations of the basic synchronism (**sync.**) system can be provided. The timebase can be synchronized from an external source, from either the rising or falling edges of the Y input, or even a range of input signal levels, to avoid false sync. through noise or **glitches**. When a second auxiliary timebase is provided, this can be used as a delay to sweep the sync. system through sections of very long waveforms in order to examine short period intervals of interest. Again, in some oscilloscopes **pretriggering** is provided so that it is possible to examine part of a waveform that would appear before the normal sync. pulse.

Like all test instruments, it is important to be able to rely on the accuracy of the readings obtained. The CRO usually has a built in calibrator that periodically needs to be checked against some standard. The substandard typically consists of an internally generated square wave with a frequency of 1 kHz and amplitude of 1 V. When used as the working input, the gain of the Y amplifier can be preset to indicate an amplitude of 1 V. The periodic time for this waveform is 1 ms and thus the duration of one timebase scan can be similarly adjusted.

The resistive component of the input impedance will load the d.c. conditions of a circuit, while the capacitive component can distort the signal waveform to be examined. While the total input impedance at the Y sockets can be around 10M in parallel with 20pF, this can be seriously compromised by the cable used to connect to the circuit under test. The use of a low-loss coaxial cable that is too long can easily distort the leading and trailing edges of square waves. Even the sine-wave frequency response can be affected because the combination of circuit output resistance and CRO input capacitance can act as an integrator or a low-pass filter.

When the CRO is used to measure pulse waveforms in medium to high-resistance circuits, a low-capacitance **probe** should be used. Such probes are available as extras for most types of oscilloscope. A few probes are active in that they contain transistor or field-effect transistor (FET) amplifiers, but most are passive, containing only a variable capacitor and a resistor, as shown in Figure 30.3. The input voltage is divided down, so that a more sensitive voltage range must be selected; but the effect of capacitance is greatly reduced because the capacitance of both the cable and the CRO input is used as part of the divider chain.

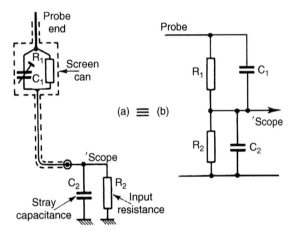

Figure 30.3 Passive probe circuit: (a) layout, and (b) equivalent circuit

Analogue oscilloscopes usually require the user to work out the effect of the probe on signal amplitude. For example, if the oscilloscope is on a 1 V per division range with a ✕10 probe then it is read as 10 V per division. Digital and personal computer (PC)-based instruments usually allow the probe factor to be entered in a menu and then correct all readings to that value.

Be careful with probes that can be set to different ranges using a switch. If the readings seem unreasonable, for example 30 V in a 5 V circuit, the oscilloscope is probably set for ✕10 probe and the probe is set for ✕1.

Oscilloscope probes must be matched with the oscilloscope input that they are used with and to this end frequency compensation presets are provided, at one or other end of the probe lead, marked (b) in Figure 30.4.

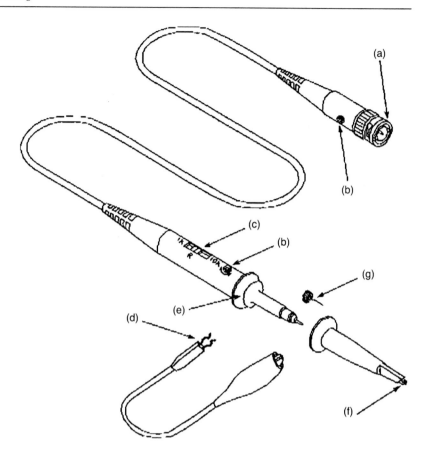

Figure 30.4   Oscilloscope probe: (a) BNC connector plugs into probe socket on oscilloscope front panel, (b) frequency compensation preset, (c) range switch, (d) ground lead, (e) ground lead connection below finger guard, (f) probe hook, and (g) spring ground connection

The compensation matches the resistive divider to the oscilloscope input, which otherwise can look like a low-pass filter. A square-wave test signal with fast rise and fall times is provided on the front panel of the oscilloscope to assist in the adjustment of the compensation preset. Probes should be checked (Figure 30.5) using this signal from time to time, and particularly if they have been used with different oscilloscopes, to check that they still match.

Oscilloscope probes are usually of the switchable type, with ×1 and ×10 ranges and sometimes a reference position selectable on the probe. The reference position grounds the probe circuit locally so that the d.c. level can be adjusted on the oscilloscope display. Most probes isolate the tip when the reference position is selected to avoid dangerous short-circuits.

In the equivalent circuit of the low-capacitance probe (b), C_2 includes the stray capacitances of the cable and of the oscilloscope. C_1 is varied until $R_1C_1 = R_2C_2$. The signal attenuation which is given by the expression

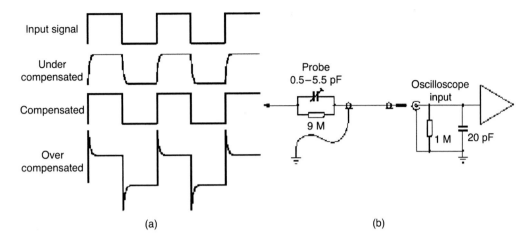

(a) (b)

Figure 30.5 Oscilloscope probe compensation: (a) waveforms, and (b) schematic of a ×10 passive probe

$R_2/(R_1 + R_2)$ shows how it is necessary to select a more sensitive oscilloscope range setting. The probe is normally calibrated against the inbuilt 1 kHz 1 V peak-to-peak (p-p) internal square-wave calibration source. The basic probe arrangement illustrated here can be extended to provide a ×1/×10 input attenuator probe by including a switched resistor network. Again, this device has to be calibrated in a similar manner to that described above.

To measure the peak-to-peak amplitude of a signal, the vertical distance, in centimetres, between the positive and negative peaks must be taken, using the graticule divisions. This distance (y in Figure 30.6) is then multiplied by the figure of sensitivity set on the **Volts/cm** input sensitivity control.

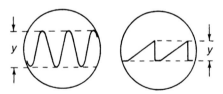

Figure 30.6 Measuring peak-to-peak amplitude

To measure the duration of a cycle of an a.c. signal, its periodic time, between successive positive or negative peaks, the horizontal distance between them is taken, using the graticule scale (Figure 30.7). This distance in centimetres is then multiplied by the time value read off the **Time/cm** switch scale. The frequency of the wave can be calculated from the formula:

$$\text{Frequency} = \frac{1}{\text{Periodic time}}$$

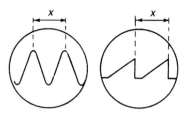

Figure 30.7   Measuring the periodic time of waveforms

With time measured in units of seconds, the calculated frequency will be given by the formula in hertz (Hz). If the time is measured in milliseconds (ms), the frequency will be in units of kilohertz (kHz); if the time is in microseconds (μs), the frequency will be in megahertz (MHz).

To use the CRO to maximum advantage, especially its more sophisticated modes of operation, it is important to obtain as much real-time practice as possible.

Practical 30.1

Connect a signal generator to the input terminals of the CRO. Set the signal generator so as to provide a 1 kHz square wave of 1 V p-p. Adjust the CRO so as to obtain a locked waveform, and read the amplitude and time. Does the time reading correspond to the frequency as set on the signal generator? Compare these settings with those obtained from the inbuilt calibration signal. Change the signal generator waveform, amplitude and frequency settings, and measure the new waveform's amplitude and periodic time on the CRO. Check that these values agree with the generator settings.

An oscilloscope fitted with an **external trigger** (EXT. TRIG.) input can be used for comparing the phases of two waves if a double-beam CRO is not available. Feeding a signal into the EXT. TRIG. input, with the trigger selector switch set to EXT, will cause the timebase to be triggered by that wave-form. The timebase will now always start at the same point in the waveform. The triggering wave can be seen on the screen by connecting the Y-INPUT socket to the same source. The X and Y shifts can be used to locate one peak of the wave over the centre of the graticule, as illustrated in Figure 30.8, which shows one leading edge of a pulse coinciding with the vertical line of the graticule.

Now the Y-INPUT from the first waveform is disconnected and replaced by the second source at the same frequency. A locked trace will appear on the screen. If there is a time difference between the two waveforms, the peak of the second wave will not be over the centre of the screen, because the timebase is still being triggered by the first waveform. Measuring the distance X horizontally from the centre allows the calculation of the time shift

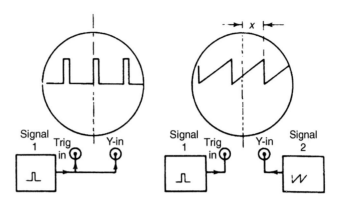

Figure 30.8   Using the external trigger

either earlier (left of centre) or later (right of centre) of the second waveform compared with the first. This time difference can be converted into phase angle if the two waveforms are sine waves. The conversion formula is:

$$\theta = \frac{360 \times t}{T}$$

where θ is the phase angle in degrees, t is the time difference, and T is the time of a complete cycle expressed in the same units as t.

Example: A complete cycle of a waveform takes 3 ms, and a second wave has its peak shifted by 0.5 ms. What is the phase difference between the two waves?

Solution: Using the formula above, $\theta = (360 \times 0.5)/3 = 60°$.

Periodically, both the X and Y amplifier circuits need to be recalibrated. Carry out this operation using the internal square-wave standard as follows.

Check the bandwidth of the Y amplifier chain by proceeding as follows. Using a standard sine-wave signal generator, provide an input signal with an amplitude to produce a 5 cm p-p trace. Progressively increase the generator output frequency while maintaining the same input level and note the frequency at which the amplitude falls to $5 \times 0.707 = 3.535$ cm (approx. 3.5 cm). This is the upper cut-off frequency, and since the Y amplifier must have a d.c. response, it is also the amplifier bandwidth. Check that this figure agrees with that stated in the instrument service manual.

Digital measurements

Processing a digital signal will create pulse distortion. What starts as a train of square pulses rapidly approaches sine waves as the high-frequency components are progressively removed. This can be measured using an eye diagram. The basic principle is shown in Figure 30.9(a), where a digital data

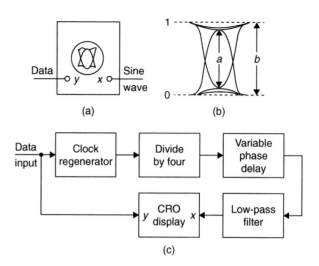

Figure 30.9   Producing eye diagrams

stream is applied to the CRO Y input. A sinusoidal timebase is provided at the external X input, running at a quarter of the bit rate.

When the phasing of the two inputs is correctly adjusted, an eye pattern appears as indicated in Figure 30.9(b). The upper and lower levels of this trace, which represent the levels produced by a run of several consecutive 1s and 0s in the data stream, are taken as the maximum eye opening. A large amplitude of eye opening represents a low level of data distortion and hence a low bit error rate (BER). The ratio of $(a/b) \times 100\%$ can therefore be used to evaluate the quality of the data signal at various points in the signal-processing chain.

Figure 30.9(c) shows the classical way in which an eye height display can be produced. The original data stream is used first to lock a local clock oscillator circuit to the bit rate. This is then divided by four and low-pass filtered to provide the one-quarter rate timebase frequency. The data and sinusoid are then input to the CRO as described above. An eye pattern will be displayed when the variable phase delay is correctly adjusted.

Storage oscilloscopes

Storage oscilloscopes are used to display fast, one-shot, transient events or very low frequencies that change too slowly to be captured by the normal-persistence CRO tubes. These instruments are available in either digital or analogue form, but the analogue form is by now obsolete. In the digital form, the signal of interest is first captured, then digitized and stored in a semiconductor memory. From here it is readily available for future display and analysis.

Figure 30.10 shows the basic principles of the operation of a digital storage oscilloscope (DSO). The work signal or Y input stages are very similar in circuitry and function to those found in the conventional CRO. High input impedance, high bandwidth and fast rise time are typical of those parameters found in the best of analogue instruments. This input stage

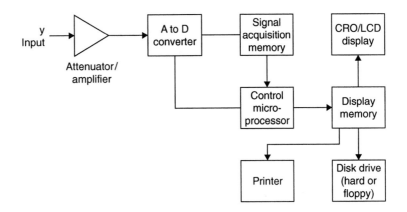

Figure 30.10   Block diagram of a digital storage oscilloscope

carries the familiar calibrated attenuator, together with buffer amplifiers to avoid loading the driving circuits under investigation. This stage is followed by analogue-to-digital signal conversion before the data is stored in a semi-conductor signal acquisition memory, all under the control of the software stored in the control microprocessor. The processed data is then passed through the microprocessor bus system to the display memory, where it is organized into a high speed raster scan format for readout and then display on a CRT or liquid crystal display (LCD) screen.

The display data can be stored in a disk memory (either hard or floppy drive) for future replay and investigation. It is also possible to obtain a hard-copy printout via a standard printer port.

An advantage of digital oscilloscopes is their ability to measure the input signal characteristics directly, for instance peak-to-peak voltage or frequency. Figure 30.11 shows a screen capture from a digital oscilloscope (TDS3200) in which the measurements and settings of the channels are clearly displayed.

Both the analogue and digital CROs still have their place within the servicing environment because each type has advantages that are not found in the other. The analogue real-time (ART) instruments have been in use for a long time and their operation and use are thus very well understood. Because the display is in real time, any fleeting signals become very difficult to capture by any recording method, other than a camera. This in itself would create film processing problems if the Polaroid Land camera had not been developed for this purpose. Even with this it is still very difficult to synchronize and photograph a very brief fleeting signal feature unless it is in some way repetitive. The DSO can display and store slow waveforms and short duration transients with equal ease. It also has the ability to store the waveform data and rerun it on demand. The DSO can perform watchdog facilities such as searching for a glitch or other fleeting signal feature over a long period, plus providing multichannel operation, with automated measurement facilities. The ART provides waveform information that is truly real time, while the DSO outputs are delayed somewhat owing to signal processing.

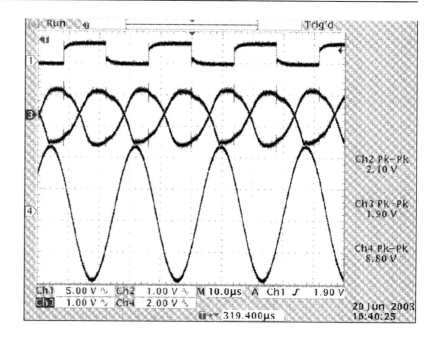

Figure 30.11 Typical oscilloscope display, showing channel amplitude settings and measurements

The input signal must be sampled at a rate at least twice that of its highest frequency component to avoid aliasing, which creates distortion and introduces additional frequencies into the waveform. The quantization rate must be at least 8 bits and preferably 12 bits per sample. For example, if the DSO has a response extending from d.c. to 500 MHz, then the sampling rate must be at least 1 GS/s (giga samples per second). At 10 bits per sample this represents a bit rate of 10 Gbit/s. Because of these features, the DSO has a greater accuracy for both the Y system and the X timebase.

A complex signal with many components, such as that found in television systems, is not always easy to resolve using either an ART or a DSO.

Because of the extensive add-ons that have been introduced into both systems over time, it has become rather difficult to learn how to manage such instruments effectively with a wide range of features. Practice therefore is particularly important.

Practical 30.2

Use digital oscilloscopes (a) in real-time mode, and (b) in storage mode. A PC virtual oscilloscope can be used if dedicated hardware is not available.

By the very nature of the instrument, the CRO carries a great deal of information about its serviceability within its display. Hence, by observation of the screen display, many of the typical faults may be localized to one of three separate areas:

- faults that affect the CRT and the power supplies
- faults associated with the X input and timebase stages
- faults in the Y amplifier stages that affect the work signal being displayed.

Individual faults can then be located using a second CRO.

The procedure for replacing a faulty or damaged tube varies somewhat from one oscilloscope to another, and the manufacturer's handbook should always be consulted. The glassware is most vulnerable at the tube neck and base connector where the glass changes shape most rapidly. The following general notes will nevertheless be useful.

1. Always wear safety goggles and gloves to avoid injury from flying glass; there is a the risk of implosion.
2. Cover the workbench surface with a blanket or similar material. Place the CRO on a clean bench, with sufficient space available to lay the CRT alongside it once it has been removed.
3. Disconnect the instrument from the power supply, and discharge all capacitors. Any outer coating of carbon on the tube should be treated with respect; it may have acquired a significant charge which needs to be earthed for safety.
4. Remove carefully the tube base, the high-voltage connectors and the leads to the deflection system. For a circular tube it will be necessary to note the particular alignment in respect of the deflection connectors.
5. Carefully remove the tube clamps and screens. Lift the tube out carefully to protect it from any impact. Place the tube on the blanket so that it cannot roll off on to the floor (circular tubes only).
6. Observe all safety precautions until the displaced tube is safely packed into its crate for disposal and then complete the reassembly of the CRO.
7. Finally, test and recalibrate.

Multiple-choice revision questions

30.1. The graph, right, shows a waveform on the screen of a CRO with the timebase set to 1 ms/cm and the Y attenuator to 1 V/cm. The sawtooth wave is of:
(a) 2 V p-p and 500 Hz frequency
(b) 2 V p-p and 2 kHz frequency
(c) 1 V p-p and 2 kHz frequency
(d) 1 V p-p and 500 Hz frequency.

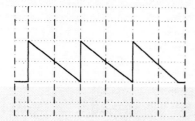

30.2. The graph, right, shows the waveforms on the screen of a CRO with the timebase set to 10 ms/cm and the Y attenuator to 5 V/cm. The frequency and peak-to-peak amplitude of waveform A are:
(a) 50 Hz and 10 V
(b) 20 Hz and 20 V
(c) 20 Hz and 10 V
(d) 50 Hz and 20 V.

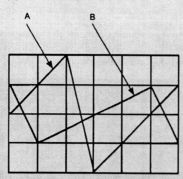

30.3. In the same graph with the same settings, waveform A leads waveform B by:
(a) 20 ms
(b) 270°
(c) 90°
(d) 30 ms.

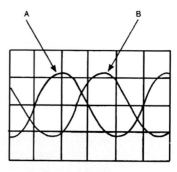

30.4. In the same graph with the same settings, the frequency and periodic time of both waveforms are:
(a) 50 ms and 20 Hz
(b) 20 ms and 50 Hz
(c) 20 ms and 20 Hz
(d) 50 ms and 50 Hz.

30.5. For the waveforms shown, right, if the X timebase is set to 100 μs/cm and the Y attenuator to 5 mV/cm, the frequency and peak-to-peak amplitudes of the waves are:
(a) 400 Hz and 2.5 mV
(b) 2.5 kHz and 12.5 mV
(c) 400 Hz and 12.5 mV
(d) 2.5 kHz and 2.5 mV.

30.6. In the same graph as in question 30.5, waveform B leads waveform A by:
(a) 100 μs
(b) 400 μs
(c) −135°
(d) 135°.

31 Test and measurement

Semiconductor devices were described in Chapters 7, 8 (Level-2 Book) and 28, and you should revise these topics before tackling this chapter. The most useful workshop instrument, the oscilloscope, has been outlined in Chapter 30.

Transistor testing

From the description of a bipolar junction transistor (BJT), we can see that the first line of testing can use the measurement of the forward and reverse biased resistance for diodes. Using an ohmmeter connected alternatively between collector and base and then between base and emitter, a high or low resistance should be found when reversing the polarity of the test leads. A similar test between emitter and collector will always find one diode forward biased and the other reverse biased, so a fairly high value of resistance should be found with each polarity of test. However, this limited test will only indicate that there are no short- or open-circuits within the device. A rough measurement of gain (h_{fe}) will check the gain of the BJT.

Practical 31.1

Measure the h_{fe} of some sample transistors using a simple tester such as the circuit of Figure 31.1.

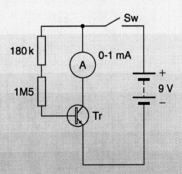

Figure 31.1   A simple h_{fe} tester circuit for BJTs. The base current will be $0.5\,\mu A$, so that a $0.5\,mA$ reading on the meter will correspond to an h_{fe} value of 100

By comparison, the metal oxide semiconductor field-effect transistor (MOSFET) is fabricated around a current-carrying channel connected between the source and drain electrodes. Control of the unipolar current flow (meaning that it consists of either electrons or holes only) is obtained by way of a capacitive effect between a gate electrode and the base substrate. Modern MOSFETs have protection diodes incorporated, so that it is possible to make meter readings that provide some indications of the condition of the MOSFET. This is not applicable to a MOSFET integrated circuit (IC), however. The simplest test is to switch the multimeter to the diode testing range and clip the negative lead to the MOSFET source and the positive to the drain. This should read an infinite resistance, no current passing. Placing the positive lead now on to the gate and then switching it back to the drain should show conductivity (because the gate retains the charge), and discharging the gate (one finger on the source and another on the gate) will drain the charge and the source-drain resistance will again be infinite.

Causes of failure

Just as diodes can become open- or short-circuit, so similar faults can arise with BJTs between base and collector and base and emitter. Individual electrodes can become open-circuit and a condition known as **punch-through** can occur where a short-circuit develops between collector and emitter when the base region is ruptured, usually through excessive current and hence heat dissipation. Other faults that may be encountered are changes in input and output impedances and gain. In general, these are due to overdriving and the consequent dissipation of excessive heat.

For field-effect transistors (FETs) the insulation layer of silicon dioxide between the gate and substrate is the major weakness. The gate region is susceptible to damage through electrostatic discharges (ESDs) and, for this reason, this electrode is often protected by diodes built into the structure during fabrication to provide a shunt path for such discharges.

Figure 31.2 shows the output characteristic for a BJT, and it is also fairly representative of a power MOSFET. At a zero value of base current I_b, the collector current I_c is practically zero for all normal values of collector voltage V_c. This is described as the **cut-off** region. As I_b and V_c increase so does I_c, as shown, and this represents the active region that is used for linear applications. At low values of V_c the collector current lies in the **saturated**

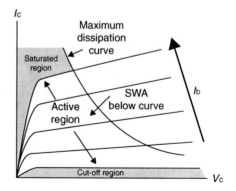

Figure 31.2   Output characteristic for a BJT showing the various areas of operation

region irrespective of the values for I_b. Since both saturation and cut-off regions represent areas of either practically zero voltage or current, the transistor dissipates very little heat and these are the regions used for digital operations. However, it is important that the transition between on and off should be rapid to avoid generating heat and wasting power source energy. This diagram also shows a maximum dissipation curve that describes the boundary of the safe working level of power [safe operating area (SOA) or safe working area (SWA) that the device can dissipate. It is a curve drawn for all values of maximum power dissipation = $V_c \times I_c$.

Semiconductor devices are relatively easy to test when out of circuit. However, when included on a populated circuit board, the dynamic effect of any signals and the shunting effect of bias and other components give rise to misleading results.

Many modern digital voltmeters (**DVMs**) now include sockets for testing the gain (h_{fe}) of small signal transistors. However, for other devices such as power transistors, although the basic circuit can be adapted it must be remembered that while small signal transistors usually have a gain that is measured in hundreds, the gain of larger power devices will be in the range of tens.

Dedicated transistor testers are available that can measure the parameters while the devices are in-circuit. In use, the circuit carrying the transistor to be tested is switched off, the test unit is connected to the transistor via short, low-resistance leads, and the residual effect of parallel circuit components is backed off with a balance control. Operating a push to test switch causes the meter to indicate the value of h_{fe}. More sophisticated test instruments are available that can also measure the other important device parameters such as input and output impedances. See the website of Peak Electronics Design (http://www.peakelec.co.uk/acatalog/jz_instruments. html) for details of useful test instruments for semiconductors and other electronics components.

Practical 31.2

From a selection of BJT devices, some of which may be faulty, carry out forward and reverse resistance tests on the three termination pairs. Record the findings and compare the results with further tests carried out using a proprietary transistor tester.

Passive component testing

All discrete passive components suffer from ageing and heating effects and, since the latter are particularly cumulative, it is important to avoid excessive temperatures during construction and repair. Furthermore, it is important to avoid the hazards of solder fumes and the environmental effects of the lead content of solder. Typically, a tin/lead solder with a melting point of about 183°C has been commonly used in the past, but this must now be replaced by lead-free alloys obtained from metals such as tin, copper, silver, indium and zinc. Like active devices, R, L, and C components are all more difficult to test in-circuit than as individual items. Hence, the simple solution is often to unsolder one end of the suspect device. All of these devices can be tested using a universal bridge tester.

Resistors very rarely develop low resistance values in use. They commonly go high in value or develop an open-circuit due to age or through passing excessive current and hence heat dissipation. Although front-line testing may usefully use a digital multimeter (DMM) switched to the ohms range, for the most accurate assessment of change of value, a Wheatstone bridge tester is by far the most effective.

Inductors (now less commonly used) may develop open-circuits through passing excessive current or simply by corrosion forming at the terminations. More commonly, inductors, like transformers, develop short-circuited turns which dissipate energy. This generates excessive heat and generally adds to the lossiness of the component. Even a short-circuit between two adjacent turns can lead to the faulty operation of any circuit to which the inductor is connected.

Capacitors are particularly sensitive to temperature effects and can develop into open- or short-circuits, change value or develop excessive leakage across the dielectric. Electrolytic capacitors are most susceptible to this problem. The important test for an electrolytic is the effective series resistance (**ESR**), which can be measured on an instrument similar to a DMM (ESR can even be included as one of the special DMM functions). In general, ESR is very low, usually less than $1\,\Omega$, and represents the loss in phase angle between the current and applied voltage, which in theory should be $90°$. Capacitors are commonly rated for operations at 85, 105 or $125°C$, so that if repeated failures occur, the replacement capacitor temperature rating can usefully be upgraded. Faulty capacitors, in general, are the most common source of circuit problems.

The basic Wheatstone bridge circuit which was introduced in Chapter 29 to determine the values of unknown resistances can be extended to evaluate unknown inductors and capacitors. In this case, the bridge must be energized from an a.c. source, the unknown device must be compared with a similar standard component, and the null detector must include a suitable a.c. rectifier circuit.

The basic conditions for the bridge circuit at balance are that the products of opposite arms are equal. In the universal bridge, the product of the range and balance resistor values must be equal, at balance, to the product of the unknown impedance and the standard impedance.

The simple bridge can only determine the component value and not evaluate its losses. In general, it is also necessary to determine the value of the equivalent series resistance of an inductor and the equivalent shunt resistance of a capacitor. In both cases, the resistive element represents the component losses which are related to its quality factor; the Q factor in the case of the inductor, and the loss angle (the angle by which the voltage/current relationship fails to reach the theoretical $90°$) for the capacitor.

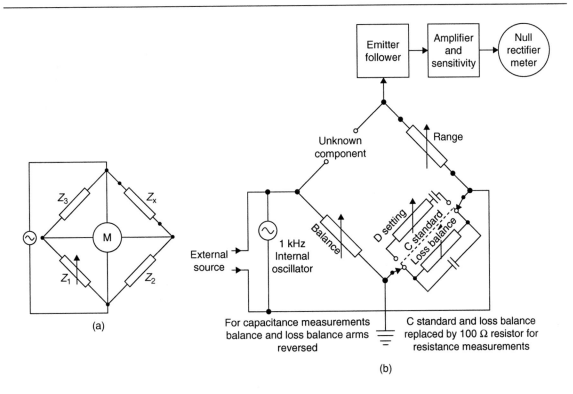

Figure 31.3 Universal bridge [courtesy of Marconi Instruments Ltd]

Figure 31.3 shows how the basic bridge concept can be adapted to provide universal features. This bridge circuit may be energized either from an internal 1 kHz oscillator or from an external source. Using an external 10 kHz source is particularly useful for measuring components with very low losses.

The null detector circuit and its amplifier and sensitivity control are buffered by an emitter-follower circuit to minimize loading on the bridge network. In application, the unknown component is connected into the appropriate arm of the bridge. The instrument is then set to measure an R, L or C component and a range is selected that provides a minimum meter deflection. A lower minimum value is then obtained by adjustment of the balance control and this indicates the component value. By further adjustment of the loss balance control, a true zero meter indication can be achieved. The loss balance setting then indicates the resistive loss component. The range of measurements available with this instrument, with an accuracy of better than ± 1% of the reading obtained, is indicated here.

Certain of these instruments are designed for portable fieldwork and are therefore likely to be subjected to shock and vibration. Local service work is therefore likely to involve the repair of such ravages. Since the overall accuracy is critical, any repair work must be followed by recalibration. The necessary equipment for this is only likely to be found in specially set-up service departments.

Practical 31.3

Using a universal bridge, compare the values obtained for a number of R, L and C components each with nominally the same values. In particular, inductors and capacitors can show a marked difference in loss values. It can also be useful to measure the self-inductance of some wire-wound resistors.

Counter/timer

These devices, which range from simple handheld models to complex laboratory bench instruments, are chiefly used for measuring signal frequency (or pulse-repetition frequency) over a range, typically from d.c. to more than 2 GHz.

The operation is based on pulse-counting over a known period, from which the frequency is automatically determined. The readout is then commonly presented on a seven-segment liquid crystal display (LCD) or light-emitting diode (LED) display. The block diagram of the basic system, omitting the reset and synchronizing sections, is shown in Figure 31.4.

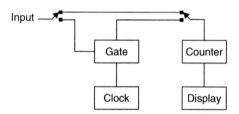

Figure 31.4   Basic block diagram of a counter/timer

The counter section consists of a binary counter with as many stages as required for the maximum count value. The input to the counter is switched, and in the counting position, the input pulses operate the counter directly. When the switch is set to the timing position, the input pulses are used to gate the clock pulses, and it is the clock pulses that are counted.

Counter/timers are often microprocessor controlled with a basic clock frequency, typically of 1 MHz in older units and higher frequencies in modern units. This is provided either by a temperature-compensated crystal oscillator (TCXO) or by an oven-controlled crystal oscillator (OCXO). Provision is often made for two-channel (A and B) measurements and these inputs may have different upper cut-off frequencies. Low-frequency noise can be troublesome with these instruments and so a low-pass filter with a cut-off frequency at about 10 kHz is often used at the inputs.

Facilities include absolute frequency, phase, period and time measurements, with frequency and time ratios for the A/B channels. Selectable gate times are used to extend the range of measurement. To extend the frequency range even further, a prescaler (a frequency divider ×10, ×100) can be connected between the source and the counter/timer. Typical sensitivities range from about 20 mV at 5 Hz to about 150 mV at 150 MHz, with input impedances of the order of 1 M in parallel with 40 pF. As for the cathode-ray oscilloscope (CRO), ×10 probes are usually available for use with signals of an amplitude greater than the in-built attenuator can handle. The resolution (the smallest change that can be registered) varies from about 0.1 Hz at the lower frequencies to 1 kHz at high frequency.

Microprocessor-controlled instruments allow for the inclusion of more elaborate features, such as hard-copy printouts and links to automated test equipment (ATE) networks.

Apart from power-supply problems, all the fault tracing will use digital techniques. Again, it is important that any service work should be concluded by recalibration, therefore this form of work should only be entrusted to suitably equipped service departments.

Frequency meter

The frequency meter makes use of a high-stability crystal-controlled oscillator to provide clock pulses. The instrument may either be combined with a counter/timer or provided as a function of a complex DMM, in which case, the accuracy will be somewhat reduced. In its simplest form (Figure 31.5), the unknown frequency is used to open a gate for the clock pulses.

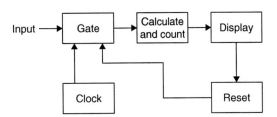

Figure 31.5   Block diagram of a basic frequency counter

The number of clock pulses passing through this gate in the period of one input cycle will provide a measure of the input frequency as a fraction of the clock rate. For example, if the clock rate is 10 MHz and 25 cycles of the unknown frequency pass the gate, then the input frequency is 10/25 MHz, which is 400 kHz. A calculating and counter-circuit then passes this result to the display stage to give a direct indication of the input frequency.

In this simple form, the frequency meter cannot cope with input frequencies that are greater than the clock rate, or with input frequencies that would be irregular submultiples. For example, it cannot cope with 3.7 gated pulses in the time of a clock cycle. Both of these problems can be solved by more advanced designs.

The problem of high frequencies can be resolved by using a switch for the master frequencies that allows harmonics of the crystal oscillator to be used. The problem of difficult multiples can be solved by counting both the master clock pulses and the input pulses, and operating the gate only when an input pulse and a clock pulse coincide. The frequency can then be found using the ratio of the number of clock pulses to the number of input pulses.

Suppose, for example, that the gate opened for seven input pulses and in this time passed 24 clock pulses at 10 MHz. The unknown frequency is then $10 \times 7/24$ MHz, which is 2.9166 MHz. Frequency meters can be as precise as their master clock, so that the crystal control of the master oscillator determines the precision of measurements.

The instrument is useful for checking and adjusting a wide range of oscillator settings, crystal oscillators in frequency conversion stages, display driver clocks and motor speed controllers.

The readout should extend to at least eight and preferably 10 digits, with an accuracy of at least 2 ppm (parts per million: 2 in 10^6) and a sensitivity of at least 10 mV. A prescaler can be used to extend the range of frequency operation, and preselectable gate times of 0.1, 1 and 10 s are often provided, but with progressively longer settling times.

For recalibration purposes, a service accessible preset adjustment provides a means of checking the master oscillator setting against some frequency standard. In the absence of a suitable laboratory standard use can be made of the 50 Hz and 15.625 kHz field and line frequencies from a television receiver, the broadcast colour television subcarrier at 4.43361875 MHz or even the BBC long-wave programme at 198 kHz, whose carrier is maintained to an international frequency standard.

Low-frequency signal generator

Some equipment cannot be tested without an input signal. Very often a signal generator will make measurements easier than the sensor or system that is used in actual operation. Applying a single frequency at controlled amplitude to an audio amplifier, for example, allows measurement of gain, and changing the frequency can explore the bandwidth.

By comparison with other signal sources, these instruments are relatively simple devices, but if they are to be trusted they must be well designed and constructed. As shown in Figure 31.6, the fundamental frequency is generated typically by a Wien bridge oscillator stage because of its wide basic frequency span. This stage is followed by a buffer amplifier to isolate the oscillator from the effects of any load. The output signal level is controlled by an attenuator stage that delivers an output typically at 600 Ω impedance.

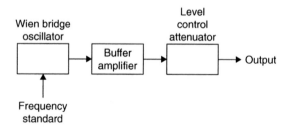

Figure 31.6　Block diagram of a low-frequency signal generator

Low-frequency signal generators are usually used for the setting up and servicing of audio systems and as such they produce output signals of both sine- and square-wave form. The frequency range usually covers from 10 Hz to 1 MHz in five switched decade bands, with a frequency accuracy ranging from 1% to 5% at full scale.

Because of the needs of hi-fi systems, these generators usually provide a very low level of sine-wave distortion, typically better than 0.05% over the range 500 Hz to 50 kHz. Off-load outputs of 20 V peak-to-peak (p-p) are common, with the level being controlled in a series of steps of −10 or −20 dB, down to about 200 mV minimum.

Practical 31.4

Measure the stage gains of a three-stage audio amplifier using the controlled output of a low-frequency signal generator, together with a suitable instrument to measure the signal amplitudes. Evaluate the −3 dB cut-off frequencies to determine the bandwidth for the complete amplifier.

Function generators

Basically, these are signal generators designed to produce sine, square or triangular waves as outputs. Each may be varied typically over a frequency range covering from less than 1 Hz to about 20 MHz. The basic frequency may be generated either by a highly stable oscillator circuit, or by using a frequency synthesizer. The output levels are typically variable between about 5 mV and 20 V peak of either polarity, plus a transistor–transistor logic (TTL)-compatible signal at ±5 V. Some of the basic waveforms are shown in Figure 31.7. The output impedance is nominally 600R and provision is made for driving signals via balanced or unbalanced lines.

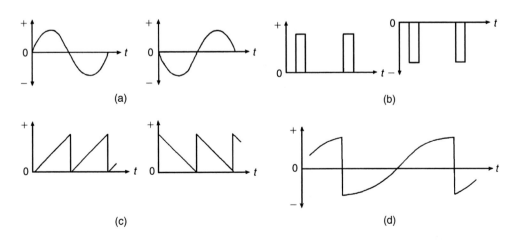

Figure 31.7 Function generator waveforms: (a) sinusoidal, (b) positive and negative pulses, (c) positive and negative sawtooths (**serrasoids**), and (d) compound wave

Each of the basic output signals is capable of being modified. The sine wave may often be phase-shifted, and the square wave mark-to-space ratio is variable so that a pulse stream of variable-duty cycles can be provided. The triangular wave can be varied to provide a sawtooth of varying rise and fall periods. In addition, it is also possible to add a d.c. value to each output as an offset. This is valuable for testing circuits that are d.c. coupled and with a frequency response that extends down to zero. The square wave is also differentiated and integrated to generate exponential envelopes (see the block diagram of Figure 31.8).

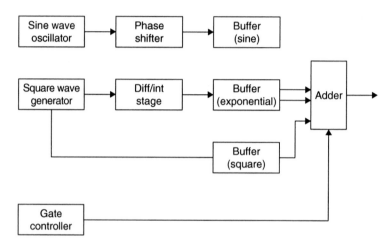

Figure 31.8   Block diagram of function generator

An additional feature often allows a second signal component to be gated into the primary waveform, as indicated in Figure 31.7(d). A further variation of this is the tone burst that consists of a sine wave modulated with a parabolic envelope, but generated in bursts. This waveform is particularly suited to the testing of full-bandwidth audio systems or those circuits where strange resonances may occur. A variation of this feature uses a form of frequency modulation (FM). Here, a basic sinusoid is frequency-swept between predetermined limits, at either a linear or a logarithmic rate.

The more sophisticated instruments in this range may be equipped for microprocessor control, and have a digital readout of frequency and amplitude, plus a built-in standard frequency source.

It is most important that the instrument calibration is carefully checked following any repairs. Service work is therefore a very specialized operation. Integrated circuits are extensively used, together with stabilized switched-mode power supplies. The highly critical oscillator circuit is invariably temperature controlled.

Decibel measurements

Measurements of gain or loss scaled in decibels can be made using an intelligent DMM. An intelligent DMM allows for automatic selection of ranges and for true root mean square (r.m.s.) measurements, so that amplitude measurements can be made. Measurements of power loss and gain can be carried out with such an instrument, and some can also measure signal-to-noise ratios.

The personal computer as a virtual instrument

With the introduction of the personal computer (PC) to the workshop environment, initially for information only, a wide range of hardware devices that are software controlled became available. These range from fairly simple and low-cost active probes capable of providing the actions of storage oscilloscope, spectrum analyser, DVM and even transient recorder functions, through DMM, storage scope with spectrum analyser, transient recorder, plus function generator, to high-cost extensive networked systems that provide all of the above, plus remote data logging from test benches, etc.

In general, these all have one point in common: they are attached to the PC via the parallel printer port, a specially provided plug in parallel port card or, more recently, the universal serial bus (USB) connector. For oscilloscope simulation the sampling rate of analogue-to-digital conversion (in MS/s) must be at least twice the bandwidth for sine waves and preferably higher (up to eight times the bandwidth) for a complex signal.

Although there are other suppliers of virtual instrument (VI) equipment, the following devices and systems that run in the familiar Windows environment, with typical task bars and pull-down menus, have been extensively evaluated for the servicing environment. These are manufactured by:

- National Instruments Corporation (UK) Ltd (www.ni.com/uk)
- Pico Technology Ltd (www.picotech.com)
- TiePie Engineering (UK) (www.tiepie.com)

Of the three manufacturers, National Instruments (NI) probably provides the most extensive range of measurement, data acquisition (DAQ), data logging including image acquisition, and signal conditioning accessories for a system that is more likely to be used in a product life testing and monitoring environment. Since the system is designed to operate with real-time applications, in a measurement, control and statistical analysis mode, it is capable of being coupled into a wide range of communications networks that include back-plane bus extensions, Internet, general-purpose interface bus (GPIB), ethernet, firewire (IEEE1394) and USB.

Standard function generator waveforms including composite video are provided, plus many more that are user programmable. These offer:

- sine and square waveforms at frequencies up to 105 MHz
- linear and logarithmic frequency sweeps and bursts
- multidevice synchronization for channel expansion
- frequency resolution up to 1.07 μHz using direct digital synthesis.

The DMMs have 5.5 digit accuracy for a.c. voltage and current (true r.m.s.), d.c. voltage and current, and resistance. These are used to measure the outputs from many different types of transducer on several channels virtually simultaneously.

The data-capture analogue-to-digital converters (ADCs) or digitizers can be driven from a wide range of transducers via appropriate signal conditioners, ranging from resistance temperature detectors (RTDs) to thermocouples, thermistors, chromatography sensors, strain gauges, force, load and pressure sensors, and linear displacement devices such as the linear variable differential transformer (LVDT). Bandwidths vary from 100 MHz down to 4 MHz, while the corresponding resolution ranges from 8 bits up to 21 bits.

Because of the complex nature of the NI system, the manufacturer provides extensive technical backup comprising system development support, technical training and future development. The website also offers training videos online, which are very useful in deciding how to make use of the products. For lower level servicing applications, however, the NI equipment may be seen as over the top for a small workshop.

Practical 31.5

The waveform shown in Figure 31.9 can be used to derive the relationship between the peak and r.m.s. value for a symmetrical square wave as follows. Squaring the voltage waveform that swings between -0.4 and $+0.4$ V produces a new wave for $V2$ as follows:

$$(-0.4)2 = +0.16 \text{ and } (+0.4)2 = +0.16$$

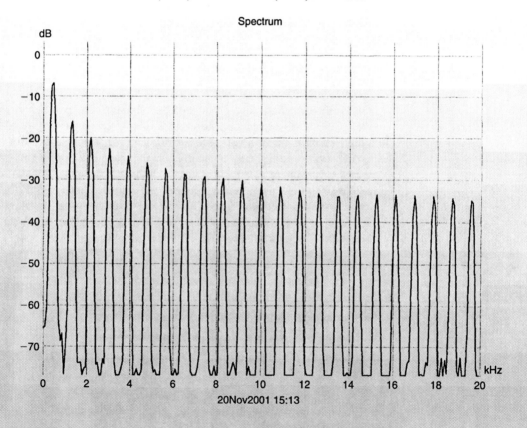

Figure 31.9   CRO display of square wave signal in time domain

(Continued)

Practical 31.5 (Continued)

The average value for $V2$ is thus 0.16 and continuously positive. Taking the square root of $V2$ produces the values of 0.4 V, again continuously positive. Thus, for the symmetrical square wave the peak and r.m.s. values are equal. This is only true for square waves; all other shapes have a variation in this relationship. Now suppose that the addition of a d.c. offset produces an asymmetrical square waveform that swings between -0.2 and $+0.6$ V (amplitude is unchanged). Repeat this exercise and compare the results.

The Pico system provides similar types of function, but with lower accuracy and less sophistication and at lower cost than the NI system. This is an economical alternative to standard test equipment and data acquisition tools. The system software and documentation are supplied in the most European languages, with a backup service via the Internet and website. The system is fully developed to CE under ISO9001 standards. It functions on most of the early PC developments from DOS software and 80386 or higher processors supported by downloadable updates from the website. Some adapters plug into the PC parallel, others into the USB connector, and the power supply of an adapter is taken from the port; no separate supply is needed.

In the time domain, the main screen of the oscilloscope provides for both normal timebase and X–Y operations. The latter is particularly useful when comparing two time-related waveforms such as found with advanced modulation schemes. The spectrum analyser provides an alternative view of a signal in the frequency domain (Figures 31.9 and 31.10). Switchable probes of 110 or 10/100 are available, in either 60 MHz or 250 MHz form.

In the digital storage CRO mode, the sampling speeds depend on the model that has been selected. Pico offers oscilloscope adapters in ranges described, as simple, general-purpose and high-speed, with sampling rates ranging from 20 to 200 MS/s, with comparable analogue bandwidths. Both waveforms and instrument settings can be stored on disk for future study and analysis.

When equipped with suitable sensors/transducers, the data acquisition and logging software can measure temperature, humidity, sound pressure, light, current, resistance, power, speed and vibration levels. This data can be transferred to other Windows applications over local area networks or the Internet using the copy and paste facility. The data can also be displayed in a spreadsheet format, such as Excel, with analysis of data trends.

The Pico handheld oscilloscope (Figure 31.11) is particularly suitable for field service (used along with a laptop computer). The large probe is plugged into a USB port, needing no other source of power, and requires no additional connections or installation procedures. The design allows for single-handed operation, with control accomplished by a button which is pressed to start the oscilloscope action and which flashes green to indicate that the oscilloscope is active. The tip of the probe is illuminated to make

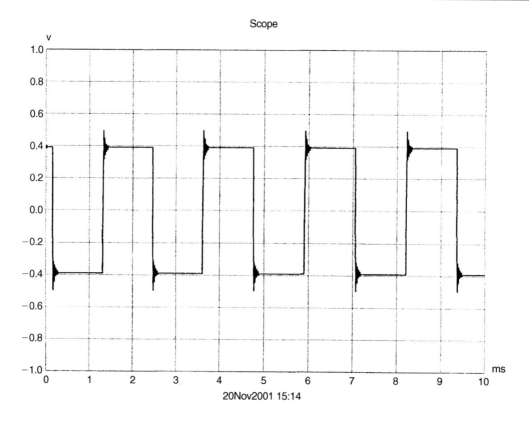

Figure 31.10 CRO display of square wave in frequency domain

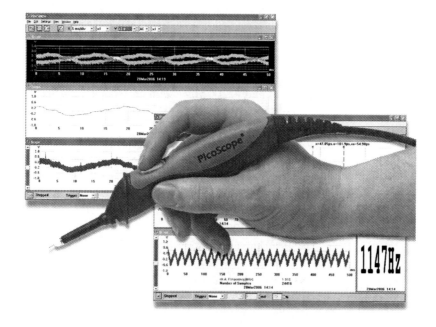

Figure 31.11 Pico handheld oscilloscope [photograph: picotech.com]

it easier to locate the point at which the signal is to be examined. When the signal has been captured on the computer screen the button can be pressed again to stop oscilloscope action: the button then glows red. If the button is held down it will activate the auto-setup action to configure timebase and triggering.

The supplied software will provide oscilloscope, spectrum analyser and metering functions, and additional software can be bought to add data acquisition.

Practical 31.6

With a PC virtual instrument, connect the system to a working three-stage low-frequency amplifier to measure the frequency response, bandwidth and gain of the individual stages. Using the spectrum analyser and an input from the function generator set to 100 Hz sine and square waves, note the change in the harmonic content at the amplifier output.

Servicing information

With the introduction of computing systems, the lifetime of the hard-copy service manual is nearly over. Much of today's service information is now provided either on CD-ROM or via the Internet. In earlier times, the service technician's manual carried not only the original service diagram and data, but also a lot of personal information that he or she had gleaned over time about that particular system. Trying to read today's manual from a monitor screen is certainly a different experience. Following the diagram of such as a colour television receiver, with interactions across many screens, appears at first sight virtually an impossibility. Even printing out sections of the diagram and gluing them together with sticky tape is not conducive to best use of the servicing bench. Fortunately, most electronic-based service information provides space in memory for the technician to add further comments as aids for future use. However the service information is provided, via the web, CD, VCR tapes, DVD or even satellite links, new skills will have to be developed to make the most of what could be information overload.

Technicians servicing domestic electronic equipment are generally well provided with circuit diagrams and data sheets, but to produce a thoroughly reliable and economical repair that will generate customer loyalty, they need additional skills and support. Not only is it necessary to understand how the system works, there is also a need to have a knowledge of the system's historical reliability and its particular points of failure. In the past, when systems were constructed largely from discrete components, many of the failures occurred in a regular manner, giving rise to the stock faults. Indeed, many service departments found these to be a source of good business that created a sound reputation for doing a good job.

With today's extensive use of dedicated ICs the system reliability has improved considerably. However, stock faults still occur, but repeat very much less frequently, making a good memory an additional requirement for service personnel. One of the best service aids is a subscription to the journal *Television* (see below). This acts as a clearing house for the hints, tips and solutions to problems encountered by many practising service technicians. Fortunately, there are now a number of computer system databases available that have been designed to aid the servicing of such equipment. While these can provide almost instant access to many of the stock faults, it must be emphasized that these should be used in conjunction with the manufacturer's circuit diagrams and data sheets.

The Mauritron website (http://www.servicemanuals.co.uk/is) extremely useful for servicing information that is hard to get or out of print, and for anyone in a hurry, a manual can be downloaded for almost immediate access. The Highland Electric site (http://www.highlandelectrix.fsnet.co.uk/myweb/auds.html) is also useful for television servicing.

Of these databases, two have been tested and found to be invaluable to the busy workshop.

The first one tested was formerly provided on CD-ROM by SoftCopy Ltd and is now available from the magazine publisher Nexus Media Communications Ltd (Media House, Azalea Drive, Swanley, Kent BR8 8HU, UK). Nexus publishes, among others, the magazines *Electronics World* and *Television and Electronics Online*. For details, see the website: http://www.televisionmag.co.uk/mag/default.asp?dir = tv

The database provided for testing by EURAS International Ltd (Keynsham, Bristol BS18, UK; website: http://www.euras.com/) is comprehensive. The fault-finding and repair hints represent the knowledge gleaned from manufacturers, dealers and repair centres, and cover about 300 000 entries from more than 400 manufacturers. Unusually, the database information is covered in most European languages. Originally, information was sent out on CD-ROM, but this has now been superseded by a completely online service, for which you need to register to obtain servicing tips (although lists of stock faults can be seen without registration). The requirements for your PC are:

- IBM or 100% IBM-compatible PC
- 32-bit operating system (Microsoft Windows 95 or higher)
- graphics card *high colou*r or higher
- resolution 800 × 600 or higher
- browser: Microsoft or Netscape version 4.01 or higher
- Java Script active
- cookies allowed
- Java Engine installed and active.

Some other useful servicing-related websites for technical data, service sheets, spares and data are listed at the end of this chapter.

Practical 31.7

Compare information from several different sources in relation to a fault condition.

Further links

- Engineering information, BBC:
  - www.bbc.co.uk, www.bbc.co.uk/enginfo
- NTL:
  - www.ntl.co.uk
- Transmitter information:
  - www.tvtap.mcmail.com (transmitter alignment program)
- Service support database:
  - EURAS International Ltd, www.euras@euras.co.uk
- Service information forum:
  - www.E-repair.co.uk
  - www.repairworld.com
- Specialist repair services:
  - www.mces.co.uk (digital repairs)
- Test equipment:
  - www.vanndraper.co.uk,
  - www.cooke-int.com
- Manufacturers' services:
  - www.philips.com
  - www.ti.com (Texas Instruments)
  - http://rswww.com (RS Components)

Multiple-choice revision questions

31.1 The safe operating area (SOA) for a transistor is located:
    (a) at the lowest possible value of bias
    (b) when the transistor is nearly saturated
    (c) between the cut-off and saturation points
    (d) when the current is a maximum.

31.2 All types of passive components can be tested using:
    (a) a multimeter
    (b) a bridge type of circuit
    (c) an oscilloscope
    (d) an ohmmeter.

31.3 You would use a counter/timer typically for:
    (a) measuring the time a task took
    (b) measuring the frequency of an oscillator
    (c) measuring the number of times a fault occurred
    (d) measuring the mean time before failure.

31.4 An LF generator will probably use:
    (a) an LC circuit
    (b) a shift register
    (c) a Schmitt trigger
    (d) a Wien bridge circuit.

31.5 An advantage of using virtual instruments is:
    (a) they can make use of a spare computer
    (b) they are more precise
    (c) they can record readings as well as measuring them
    (d) they make use of software.

31.6 The main advantage of getting servicing information from the Internet is that:
    (a) it is more accurate
    (b) it is more up to date
    (c) it is faster and requires no storage
    (d) it is better illustrated.

32 Rework, repair, reliability, safety and European Union directives

From a customer or user's perspective, quality is a measure of fitness for purpose, as a measure of the reliability and the ease of use and efficiency of the product. From a manufacturing perspective, quality is often defined as a measure of how well a product or system conforms to its claimed specification. Reliability is then a measure of how long the product or system continues to meet this specification. Quality and reliability are of great interest to the end user because they directly affect their confidence that the product or system will meet their expectations. To meet both manufacturers' and customers' definitions it is usual for products to be designed in a quality-assured environment. Quality assurance is the management of the design and manufacturing processes in such a way as to build in quality. As part of quality assurance it is necessary to measure the manufacturing yield and field returns, among other parameters; these parameters can be evaluated statistically to provide the evidence necessary to maintain or improve the quality of the product.

Quality control is the measurement of product parameters in production and forms a vital part of the overall quality assurance process. It is not possible to measure all the parameters of a product, since parameters such as lifetime or failure rate are by their very nature to an extent random; however, a suitable statistical sampling scheme can give an accurate assessment of the state of a population.

Statistical data is acquired through inspection and testing, using standard sampling plans which define how many units from a batch of a given size need to be tested for the characteristics of the whole batch to be inferred. In addition, life testing of randomly selected units from production batches can expand the characteristic information to include failure rate and failure mode data.

A further important feature of quality assurance schemes is the concept of traceability; that is, identifying the source of component parts of a system by recording manufacturing history data. This has two functions. First, it allows audits to prove that all components used have been produced and tested to a recognized standard by an accredited manufacturer, for instance assuring the lead-free status of a product as required under European Union (EU) Restrictions of Substances Hazardous to Health (RoHS) legislation.

Secondly, in the event of returned field failures the failure mode can be identified, and if a particular component or part is responsible it can easily be traced back to the original manufacturing source. This can allow the batches where such parts were used to be identified and recalled if necessary. Such product recalls are not common, but when safety is put at risk, whether by exploding laptop batteries or faulty parts of a car handbrake, it is much better for consumer confidence that the faulty parts are seen to be withdrawn quickly and effectively.

Good component traceability can also help to improve production yields by identifying components and vendors that have caused problems in the past and either designing them out or designing for the worst case in original system design.

Quality assurance is also applicable to the service environment because it encompasses the procedures for handling material spares and consumables, ensuring that customers' equipment is maintained with the correct parts and materials and that returns are dealt with in an effective way.

Accelerated life testing

Normal life testing involves operating systems or components as near as possible to the end user's applications and conditions. However, if the accepted lifetime is very long, then data gathered by this technique is too time-consuming, so the test procedure has to be accelerated. For example, devices designed for intermittent operation would be run continuously, thus compressing the lifespan. Electronics components that are intended for continuous operation may be tested with an increased load or at an elevated temperature. Further environmental tests which simulate extreme conditions of temperature, humidity, pressure, shock, vibration and abnormally dirty conditions in a time-cycled way may be used.

Similar principles may be used for **soak testing** (leaving equipment running over a long period) after equipment has been repaired, but the facilities in many smaller repair depots may not run to such elaborate techniques. The use of a freezer spray or hot-air blower can be used to simulate extreme conditions.

Bathtub diagram

The time/failure rate curve is often referred to as the bathtub diagram because of its shape. The high level of early failures shown in Figure 32.1 is often described as **infant mortality**. These are usually due to manufacturing faults, deficiencies in the quality control scheme, design deficiencies, or misuse by or the inexperience of the end user. These failures are fairly quickly eradicated and the failure rate settles down to a constant, often zero level over the normal working lifetime. Towards the end of the useful service life, the failure rate starts to rise once more and this is described as the **wearout region**. Equipment may be 'burned in' by the manufacturer; that is, operated at full power and possibly elevated temperature for a defined period to reduce the possibility of infant mortality affecting the end user; this may also be done after repair, when it is usually referred to as soak testing.

Mean time between failures (MTBF) is a factor that is often quoted. For a system with a constant failure rate of, say, 0.001 per hour, it would be expected to fail and have to be serviced every 1000 h. Thus, the average

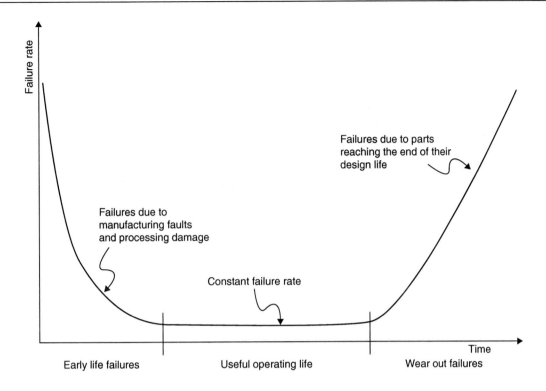

Figure 32.1   The bathtub diagram

or mean time between failures would be 1000 h. MTBF is therefore the reciprocal of the constant failure rate. When it is not practicable to replace a failed component, it is more usual to refer to the mean time to failure.

Mean time to repair (MTTR) or **downtime** is the term used to describe the length of time that a system is out of action between failure and repair.

If components and systems are operated at below their normal design parameters, their lifetime can be extended, a technique often described as **derating**. In fact capacitors often fail in a system that is running at near to the upper temperature limit for the components. Increasing the rating of a capacitor from, say, 85°C to 105°C will improve the system reliability. Typically, an electrolytic capacitor rated for service at 105°C is specified to have a service life of 6000 h at 105°C. This is just 250 days, so for most applications one would want to use it at much lower temperatures.

Even in the servicing environment, it is important to report consistent failures because this can be accumulative throughout a wide area and could lead the manufacturer to improve the design.

Surface-mounted technology

Surface-mounted technology (SMT) uses components designed to be as physically small as possible without the constraint of having to have long, stiff, wire leads attached to the component body. These small sometimes leadless components are attached by direct soldering to the metallic conductor pads on the circuit boards. One of the advantages is being smaller than their conventional counterparts, but the main advantage is that the printed circuit board (PCB) no longer requires holes for every component lead, so making the PCB both cheaper and smaller, and this leads to smaller subsystems. Because of the relatively large area of the solder joint between component termination and the solder land on the PCB, highly reliable joints are produced with good heat-dissipating properties. Usually, SMT components have lower series inductance from being physically small and not having long leads, so that these devices have better radio frequency (RF) characteristics. Surface-mounted construction is dominant in all areas of electronics, from audio to microwave systems. The technology of surface-mounted components or devices (SMCs or SMDs) is highly compatible with automated assembly, which in turn further improves the cost-effectiveness.

Most SMDs are supplied on tape and reel; typically, several thousand devices are supplied on a carrier tape, rolled on a reel much like a super 8 or 16 mm cine film reel, designed for automatic pick and place equipment. When small quantities of components are bought they are often supplied in sealed antistatic plastic bags. They should be stored in their packaging until actually needed, to avoid mixing components that are difficult to identify. It is also a good idea to keep SMCs in airtight containers to reduce the chance of the terminals tarnishing or the plastic body of integrated circuits (ICs) and some capacitors absorbing water from the air. The protection of components from absorbing water from the atmosphere and tarnishing of the connecting terminals has become particularly important since the introduction of RoHS legislation in Europe, which banned lead from solder, resulting in higher soldering temperatures and greater risk of damage to the component. It is very important to use the correct solder for the type of component and PCB being worked on: lead-based solder such as 60/40 or the fully eutectic 63/37 will continue to be available for maintenance and repair for some years to come, and these solders must not be mixed with lead-free types, as the joint reliability would be impaired.

Most discrete SMDs are too small to mark a value on them in the conventional way, with colour or laser scribing or heat-cured ink printing. Typically, ceramic SMD capacitors are unmarked, some resistors are marked usually with a three- or four-character code, inductors are similarly marked, transistors usually have a three-character code, and larger devices such as ICs and power transistors are marked in the normal way. The coding shown in Table 32.1 is commonly used.

Common resistor markings include two-, three- and four-symbol versions. Typically, the two-symbol code would be Q5 for a 390 kΩ resistor, while for the three-symbol code, the first two digits indicate the base figures and the third digit indicates the multiplier or number of zeros to add. Therefore, 270 = 27R or 27 pF, 331 = 330R or 330 pF, 472 = 4.7 K or 4.7 nF. Some manufacturers use the letter R as a decimal separator for both resistors and capacitors; for example: 2R2 = 2.2R or 2.2 pF or 2.2 μH. Four-digit codes are similar, using the first three numbers for the value and the last digit for the multiplier.

Aluminium electrolytic capacitors may use a three-symbol code with numbers to indicate capacitance value in microfarads, plus a letter to indicate

Table 32.1		SMD component marking codes									
A	1	J	2.2	S	4.7	a	2.5	y	9	6	×1M
B	1.1	K	2.4	T	5.1	b	3.5			7	×10M
C	1.2	L	2.7	U	5.6	d	4	0	×1	8	×100M
D	1.3	M	3	V	6.2	e	4.5	1	×10	9	×0.1
E	1.5	N	3.3	W	6.8	f	5	2	×100		
F	1.6	P	3.6	X	7.5	m	6	3	×1000		
G	1.8	Q	3.9	Y	8.2	n	7	4	×10k		
H	2	R	4.3	Z	9.1	t	8	5	×100k		

the voltage rating, as follows: C = 6.3 V, D = 10 V, E = 16 V, F = 25 V, G = 40 V, H = 63 V.

The position of the letter in the code indicates the decimal point in the capacitance value. For example, F47 = 0.47 μF 25 V, 3E3 = 3.3 μF 16 V, 22C = 22 μF 6.3 V.

Devices such as discrete transistors and diodes are marked according to different manufacturers' systems, so it is not unusual to find two different devices with the same code. Some common examples of transistor and diode marking for common SOT23 package devices are 1F for the Philips BC847B (which is a SMD version of the BC107B/BC547B) and A6 for the BAS16 diode.

Always check the manufacturer's data sheet when identifying SMD transistors and diodes. There are too many devices with similar codes to assume anything.

Soldering technology

Allied to the use of SMDs are the changes in soldering technology that have occurred over recent years. Repair and component replacement of surface-mounted PCBs now typically require a microscope and fine tweezers, along with a temperature-controlled iron, a hot-air pencil and a very steady hand. Great advances in reliability have been made by the use of automatic processing for surface-mounted assembly; however, it has made repair and maintenance much harder. The introduction of the RoHS legislation has also affected things because there are now several incompatible lead-free solder systems in use, as well as the legacy of lead-based equipment. It remains to be seen whether the reliability of the lead-free systems approaches that of the older, lead-based solder.

The PCBs or substrates used with SMDs often have no through-holes for conventional components; the component leadouts are soldered directly to pads or lands provided on the metal tracks. This feature required the development of new soldering and automation techniques which yield further advantages. In manufacture, the soldering methods that are used lead to improved connection reliability, which in turn leads to a reduction in costs. The particular technique used in manufacture has a bearing on the way in which SMDs can be handled during servicing.

- **Wave soldering**: the components are attached to the solder resist on each PCB, using an ultraviolet light or a heat-curing adhesive. The boards are then passed, inverted, over a wave soldering bath with the adhesive holding the components in place, while each joint is soldered.

- **Reflow soldering**: a solder paste or cream is applied to each pad on the circuit board through a silk screen, and components are accurately positioned and held in place by the viscosity of the cream. The boards are then passed through a reflow oven or over a hot plate, to reflow the solder and make each connection.

- **Vapour phase reflow soldering**: this more controlled way of operating the reflow process uses the latent heat of vaporization to melt the solder cream. The boards to be soldered are immersed in an inert vapour from a saturated solution of boiling fluorocarbon liquid, used as the heating medium. Heat is distributed quickly and evenly as the vapour condenses on the cooler board and components. The fact that the soldering temperature cannot exceed the boiling point of the liquid (215°C) is an important safety factor.

Electronic components experience distress at all elevated temperatures. For example, in wave soldering, most baths have an absolute limit of both temperature and time; normally, 260°C for no more than 4 s. As the damaging effects of heat are cumulative, manufacturing and servicing temperatures have to be kept to a minimum. The common 60/40 tin/lead solder (melting point = 188°C) and the lower melting point tin/lead alloy (fully eutectic alloy solder has a melting point of 183°C) were widely used, along with 62.5% tin, 36.5% lead, 1% silver alloy (62.5/36.5/1 Sn/Pb/Ag), which had improved leaching and vibration performance.

Components often have silver- or gold-plated leadouts to minimize contact resistance. Tin/lead solder alloys cause silver leaching; that is, over a period of time the solder absorbs silver from the component and eventually causes a high-resistance joint. This can be avoided by using a silver-loaded solder alloy such as 62% tin, 35.7% lead, 2% silver and 0.3% antimony. Such a solder has a melting point of 179°C. Tin also tends to absorb gold with a similar effect, and this is aggravated by a higher soldering temperature. This latter alloy is thus particularly suitable.

The new RoHS compatible alloys are mostly similar, 95.5% tin, 3% silver, 0.7% copper, with a high melting point (217°C). These solders should not be used with conventional non-RoHS SMDs because of the overheating risk. It is generally safer to use tin/lead solder if in doubt.

The fluxes used to clean the metal surfaces and prevent oxidation during soldering are also important. An effective flux improves the solderability of the components, the rate of solder flow and hence the speed at which an effective joint can be made. The flux used for SMD circuits should be either a halide-free no-clean compound, or one of the water-soluble types, which should be thoroughly washed away after completion of the soldering. Acid fluxes like those used by plumbers should be avoided as their residues after soldering can corrode through the thin copper PCB traces and component terminations.

Surface-mounted device values and tolerances

Resistor values and tolerances for SMDs are the same as those used in more conventional applications. Because of the block-like structure and the good heat-dissipating properties of the lead contacts, the power dissipation may be higher than expected. Typical common ratings range from 1/10 to 1/4 W. Capacitors also have a normal range of values and tolerances, except for the non-availability of very high capacitances. The common range of values extends from a few picofarads up to several microfarads, but the working voltage ranges tend to be somewhat lower. A typical ceramic capacitor range includes 50 V parts up to 100 nF and 16 V up to 470 nF. Available capacitances for a given case size fall with a rise in working voltage.

Both resistors and capacitors are also manufactured in multiple units, typically up to eight resistors or four capacitors in a single block.

Typically, inductors are available ranging from 1 nH to 1 mH at up to 5 A of saturation current. Similarly, transformers that cover frequencies ranging from audio applications to lower microwave bands at around 6 GHz are currently in use.

Packaging

The familiar single in line (SIL) and dual in line (DIL) devices that have been used for years are now supplemented by a large number of other packaging methods, particularly for computer chips and SMDs (Figure 32.2). These new packaging systems have generated new and sometimes obscure abbreviations. These include small outline package (**SOP**), with variations like thin small outline package (TSOP) and thin shrink small outline package (TSSOP), which refers to a small outline package that is smaller than standard and usually has a finer pin pitch. Discrete devices also have new abbreviations: **SOT** for small outline transistors and SOD for small outline diodes. More recently, ICs such as op-amps and voltage regulators and low pin-count microcontrollers have begun appearing in very small outline packages (**VSOP**s) or leadless laminate package (**LLP**s) and quad flat no leads (**QFN**s); these are typically the same package from different vendors made from a laminated glass epoxy material with metal contacts on the underside only.

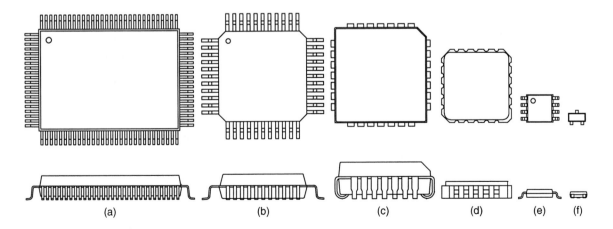

Figure 32.2   Typical surface-mounted IC packages: (a) 100-lead MQFP, (b) 44-lead PQFP, (c) 28-lead PLCC, (d) 20-pad LCC, (e) SO8, and (f) SOT23

Complex ICs may be directly mounted on to a circuit board or occasionally plugged into a socket. Large ICs with contacts arranged around four sides of a square are described as quad flat packs (**QFPs**) or plastic quad flat packs (**PQFPs**). Integrated circuits designed to be either mounted in a socket or soldered directly to the PCB use J-lead contacts and are usually described as plastic leaded chip carrier (**PLCC**). Ceramic versions, e.g. leadless ceramic chip carrier (LCC), are typically used in older high-rel equipment and sometimes for prototypes and production fixes.

A further development involves the use of ICs with several rows of contacts arranged around the four undersides of the square chip package, which terminate in sometimes more than 400 contact bumps. Each of these is soldered directly to matching pads on the circuit board by one of the methods described above. This ball grid array (**BGA**) system creates real difficulties for repair or replacement. The only way to remove the chip is to use a hot-air gun or infrared rework station and then lift off the chip once all the solder under it has melted. This needs to be done as quickly as possible to prevent damage to the surrounding components and the PCB material. For replacement purposes, pads of prealigned solder bumps are available in a grid format to be used when replacing BGA devices; generally, it is not economical to repair BGA boards.

Fault-finding on surface-mounted technology boards

As with discrete component circuit boards, the divide and conquer or **half split** technique has much to recommend itself in terms of time spent on fault-finding with SMT boards. Locating the faulty area by signal tracing is applicable to both analogue and digital circuits. Once the faulty area has been located, the component or IC will have to be tested. L, C and R devices are usually fairly straightforward. Disconnect the suspect component and check the circuit board for any unexpected short- or open-circuits associated with the contact pads that might have caused the fault. Test the displaced component to determine that it is in fact faulty before replacing with a new one. Even if doubt exists that the component might be serviceable, remember that the heating effects of removal and refitting could lead to an early failure. It is therefore often advantageous to fit a replacement.

To test a suspect IC, measure all the d.c. voltage supply points for correct value. Check for the same voltage levels on adjacent pins and check the continuity of earth lines. If signals are present at the input, and if the IC is serviceable and not overheating, then signals should be found at the outputs. However, there are a few exceptions to this rule, where some power output stages, voltage stabilizer circuits or similar may have shut down through an excessive rise in temperature.

If an IC has failed, it is important to ascertain why before fitting a new one. Check all the supply lines for open- or short-circuits while the IC is out of circuit. Check particularly for any adjacent pins that have acquired the same voltage levels. This may be caused by a short-circuit somewhere in the circuit tracks (Figure 32.3).

Servicing SMT equipment

For the larger centralized service department, a soldering rework station may be cost-effective. This may include a small portable vapour phase soldering unit, such as the Multicore Vaporette, which is particularly suited

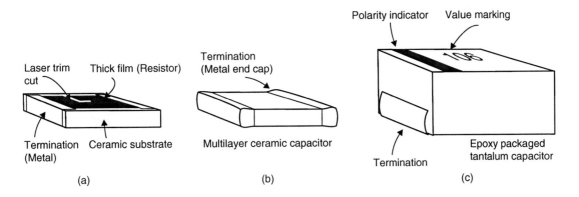

Figure 32.3 Discrete SMD components: (a) thick film resistor, (b) multilayer ceramic capacitor, and (c) tantalum capacitor

for small batch work as commonly found in such establishments. For the smaller service department, much ingenuity might be needed when dealing with SMD circuits.

The method used to remove suspect components may vary with the number of leads per device. For a device with only two or three leads, a fine soldering iron, in conjunction with a solder sucker or solder braid, can be successful. For multipin ICs, two methods are popular. One involves directly heating the soldered connections with a carefully temperature-controlled hot-air jet. The other method uses an electrically heated collet or special extension to a conventional soldering iron, to heat all pins simultaneously. When the solder flows, the component can be lifted away.

In a simpler way, all the leads can be severed using a pair of cutters with strong, fine points. The tabs can then be removed separately. This is not quite so disastrous as it appears. Any suspect component that is removed, and subsequently found not to be faulty, should not be reused; the additional two heating cycles are very likely to lead to premature failure. As the desoldering of even a few leads can be difficult, it is important that the circuit board should be firmly held in a suitable clamp.

Even after all the solder has been removed, a problem of removing the component may still exist if an adhesive was used in manufacture. Although care should be exercised when prising the component off the board, damage to the printed tracks is unlikely. The adhesive should have been applied only to the solder resist. After component removal, boards should be examined under a magnifier to check for damage to the print.

For SMD work the soldering iron used should be a temperature-controlled type rating and should not be applied to a joint for longer than necessary. If solder takes much more than 3 s to melt it may be necessary to use an iron with a greater thermal capacity. It is better to use a larger iron (temperature controlled) than a small one that fails to do the job quickly. The heat from the iron should be applied to the component termination and the PCB pad simultaneously, and the solder should flow on contact with either surface, otherwise a poor joint will result. Never rely on a joint made by carrying solder to the joint on the iron, since all the flux will have been burned away and will not clean the metal surface of oxides before the solder cools, which is necessary to make a good joint.

Before starting to fit the new components, the solder pads should be lightly pretinned, using the appropriate solder and flux, but not so much as to make the surface uneven. Each component will need to be precisely positioned and firmly held in place while the first joints are made. For multileaded components, secure two diagonally opposite leads first. For chip resistors and similar components, first apply solder to one pad (Figure 32.4a), then while heating the solder on the pad with the soldering iron to keep it molten introduce the component with tweezers (Figure 32.4b). The component should be held still until the solder has solidified, then the other end of the component may be soldered in the usual way (Figure 32.4c). The first joint may then need to be reworked with the addition of a small quantity of flux. The final joint should have a smooth continuous fillet (Figure 32.4d, A). Figure 32.4 also shows three incorrectly made joints: B is a joint with too little solder, and this produces a high-resistance connection which may also be mechanically unreliable; joint C has been made with the application of too little heat, which will also lead to a high-resistance connection; and in joint D, too much solder has been applied and this has probably resulted in excessive heating of the component.

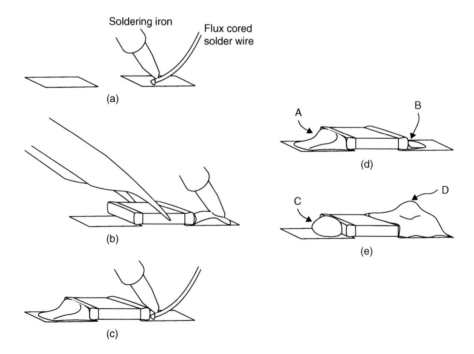

Figure 32.4 Soldering chip-type SMDs: (a) apply solder to one pad, (b) melt the solder and introduce the chip with tweezers, (c) solder the other end once the first joint has cooled, (d) visual soldering standards, a good joint at A and insufficient solder at B, (e) a dry insufficiently heated joint at C and too much solder at D

Reliability in practice

Experience suggests that the most overstressed discrete components in a system are the capacitors, particularly the electrolytics. If these are operated at too high a temperature, the electrolyte tends to dry out or leak. In the first case, this leads to a significant reduction in capacity with the attendant loss of performance. In the latter case, the leakage can create corrosion on the circuit board to generate open circuit tracks. When replacing a capacitor it is therefore important to ensure not only that the capacitance value, voltage rating and equivalent series resistance (ESR; typically $2 \times 10^{-2}\,\Omega$) are correct, but also that the temperature rating will not be exceeded. Typically, for television and satellite receiver applications, a 105°C rating is specified, but upgrading this to 125°C will often improve reliability. Again, if these devices are operated at too low a d.c. voltage, the dielectric material will start to decay. At first this causes the capacitance value to rise with the thinner dielectric, but this quickly punctures and becomes a short-circuit. Modern circuit boards tend to be liberally populated with these devices and experience has shown that when one fails the others are quite likely to follow suit in quick succession.

Safety testing

It is important to point out that there are generally two earth points within electrical equipment. The primary earth ground (**PEG**) or **protective earth** is designed to provide protection against electric shock to the user, and the secondary earth ground (**SEG**) or **signal earth** is used to provide a return path for signals within the equipment. Great care needs to be exercised where the two earths are connected together within a system. Each item of electrical equipment is connected to the mains supply via a green/yellow earth cable.

Portable appliances are pieces of equipment that are powered via flexible leads plugged into a mains single-phase electrical supply of 230 V a.c. Industrial equipment may be similarly connected to three-phase, 400 V a.c. mains, but is covered by a different class of regulation.

Within the single-phase group there are three classes of equipment, defined as follows:

- **Class 1**: with this equipment, the basic electrical insulation is additionally supported by providing a connection between the exposed metal parts and the protective earth provided by the mains power supply. This technique is used in all industrial equipment.
- **Class 2**: with this equipment, the protection does not rely on the basic insulation, but on additional internal insulation and the avoidance of any electrical connection to any exposed metal work. These are often described as being double insulated and are suitable only for domestic applications.
- **Class 3**: such equipment is designed for use with safety extra-low-voltage (SELV) supplies that operate at less than 25 V a.c. r.m.s. or 60 V d.c. Rechargeable battery-operated portable tools fall into this class.

All appliances connected to the mains supply via flexible leads and plug and socket connectors should be examined and tested at least annually by a competent person.

- Flexible leads should not be frayed, split, sharply kinked, cut or too tightly clamped.

- Damaged leads should be replaced immediately.

- Cables must be securely fastened at both the plug and the equipment.

- The supply cable to a heavy piece of equipment should be connected in such a way that the plug and socket will separate if the cable is pulled.

- The live end of the connector should have no exposed pins that may be touched.

- Every connection should be both mechanically and electrically sound.

- The cable must be well secured with no danger of working loose and with no stray strands of wire.

- Finally, be aware that hot soldering irons can easily damage power leads.

It is important to be able to test and record the earth bond resistance, earth loop impedance and insulation resistance between line, neutral and earth after any repairs. This will ensure that the work carried out meets the standards required by law.

Various suitable test units are available from manufacturers such as AVO, Megger Ltd and Seaward Ltd, which are readily portable and simple to use. The old-fashioned Megger, which is still often used, uses a hand-cranked generator to provide the test voltages, with the test results being indicated on a moving-coil meter. The modern versions are driven from a mains power supply or a 9 V battery to provide the necessary high test voltages. Most test sets have at least two ranges, 0–200M and infinity for insulation resistance and 0 to 2Ω for continuity. The indication of the results is displayed on a liquid crystal display (LCD) screen, which is supported by a battery-backed semiconductor memory. The results can then be downloaded into a personal computer (PC) memory for future reference.

The continuity and resistance of the earth lead should be tested while passing a current of 25 A for 5 s, which is high enough to open-circuit a partially fractured cable. The earth bond resistance should be in the order of 0R1 and not greater than 0R5 if the appliance plug is fused at 3 A or less. This test is carried out by inserting the appliance plug into the test set and connecting the earth test lead to its conductive casing. With this test the series resistance of the mains cable needs to be taken into consideration, but commonly this combined resistance will be less than 0R5.

The insulation resistance will require a 500 V d.c. test voltage and for an earthed appliance the live and neutral pins are shorted together and the resistance between them and the earth connection is measured. The test voltage is applied for 5 s as the insulation resistance is being measured. Typically, this should be at least 2M. For a double-insulated device that carries the double square symbol, the test lead is connected to any exposed metal work and the insulation resistance should be higher than 7M.

It is also useful to attach a self-adhesive label to the device, recording a serial number, perhaps in bar code, referring to the test data. This will help to make workers in the field aware of the continuing need to maintain this standard. Portable appliance testing (PAT) testers are available that have the facility of storing the test data so that it may be downloaded into a PC database for future recording purposes.

On occasions when it is necessary to measure the state of the electrical distribution network in, say, a small workshop, the job is only slightly more complex, but involves the following definitions.

- $R1$ = resistance of the phase conductor
- $R2$ = resistance of the protective earth conductor
- Hence, $R1 + R2$ = the loop resistance
- Zs = earth fault loop impedance
- Ze = the earthing impedance, external to the circuit
- $Zs = Ze + R1 + R2$, and generally in such an application, any stray reactance effects can be ignored.

Then proceed as follows:

1. Disconnect the network from the supply by means of the main circuit breaker.
2. Strap the phase to earth at the distribution board.
3. Test the loop and insulation resistance between phase and earth at each outlet socket.
4. Record this information.

Mains polarity

The polarity of the mains supply (not really an appropriate name because the voltage on the live line is alternately positive and negative) can be checked by using a neon lamp or probe. This device glows when in contact with a connection at a potential greater than about 100 V d.c. or a.c. Using a voltmeter referenced to earth, the live line voltage should read 230 V a.c. per phase. Theoretically, since the neutral line is connected to earth at the nearest substation, this should read 0 V. However, small induced voltages of perhaps 10 V a.c. can usually in practice be measured between neutral and earth.

Practical 32.1

Using a range of mains-powered electronic test equipment (cathode-ray oscilloscope, signal generator, electronic voltmeter, etc.), disconnect the equipment from the mains supply. With a suitable instrument or PAT tester measure and record the earth loop impedance, earth bond resistance and insulation resistance to earth.

Electromagnetic compatibility and interference

Electromagnetic compatibility (**EMC**) is defined as the ability of a device, piece of equipment or system to function satisfactorily in its electromagnetic environment without introducing intolerable electromagnetic interference (**EMI**) to any other system; at the same time, its own performance must not be impaired by interference from other sources. Such interference occurs as noise within a system and since this is considered as a destroyer of information, it is often a limiting factor in a communications system.

Interference can be divided into two categories, natural and artificial. The former often results in electrostatic discharge (**ESD**), while the latter usually results in power-line surges. Natural sources of interference include ionospheric storms and lightning. Artificial interference commonly results from high current switching operations and RF generators. With the expansion in the use of portable digital communications systems such as computers and cordless mobile telephones, sources of interference are growing rapidly.

A direct lightning strike is not necessary to cause havoc with a communications system. Even cloud-to-cloud discharges can easily set up high electric fields that can influence power lines and overhead telephone lines, which tend to act as aerials. Instances have been recorded of induced voltage spikes in the order of about 2000 V, but of very short duration, being developed on the normal 230 V a.c. supply mains. Even if exposures to such discharges do not prove to be immediately destructive, they can introduce a latency failure mechanism because these effects are cumulative.

The rubbing action between dissimilar materials has long been a recognized phenomenon, now referred to as triboelectricity. A triboelectric series lists the materials ranging from air, hands, asbestos, at the most positive, through the metals, to silicon and Teflon as the most negative. The further apart in this table, the higher will be the electric charge between the two materials when rubbed together. Even the action of a person walking across a carpet or even sitting in a chair can generate very high electrostatic charges, depending on the clothing worn and level of humidity. Owing to relationship between charge and voltage ($Q = CV$ joules), the action of a sitting person lifting their feet off the ground reduces the body capacitance and this automatically causes an increase in the voltage level to increase the risk of ESD.

Electromagnetic interference can enter communications equipment either by direct radiation or by conduction. The effects of the former can be reduced by screening or shielding, while reduction of the latter requires the use of suitable filters.

Connecting leads, particularly the longer ones, make very good aerials for propagating interference problems. Such leads are also susceptible to breaking, where they can vibrate. The loose ends then become the source of arcing or even short-circuit problems. Such leads should therefore be dressed close to the circuit boards or chassis and firmly clamped so that their capacitive pickup effect is minimized.

Mains plug filters that have a low pass characteristic, with inbuilt shunts to discharge any excessive voltage spikes to earth, are very useful for restricting the ingress of mains-borne interference, particularly in respect to digital and computing equipment. Protection for such equipment can avoid the blackout of services which lead to complete loss of system functions. Brownout is the jargon term used to describe partial loss of processed data.

- Holes in metalwork should have dimensions that are less than about one-tenth of a wavelength at the major operating frequency. A slot is a particular form of aerial that will radiate and accept signals very effectively at higher frequencies.

- All plug and socket type connectors should maintain good low-resistance contacts, because high resistance generates heat and the resistance itself then creates a noise component.
- Shielding problems can arise after servicing by failure to replace all the screws and screen fixings correctly. Because of the wavelength effect, even the spaces between the screen-securing screws can be critical in very high-frequency equipment.

The design of connections in equipment and the connections between equipment can have a big influence on how well systems operate in a noisy electromagnetic environment. One of the most common problems with equipment is incorrect grounding, causing unwanted feedback between stages and injection of signals into input stages and intermediate stages. Figure 32.5 shows how badly designed grounding can lead to problems.

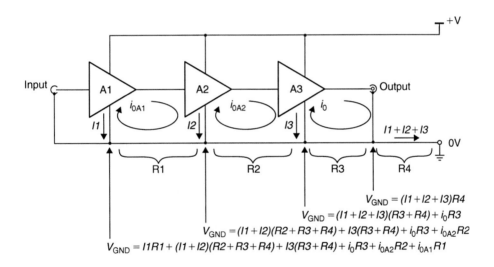

$$V_{GND} = (I1 + I2 + I3)R4$$
$$V_{GND} = (I1 + I2 + I3)(R3 + R4) + i_0 R3$$
$$V_{GND} = (I1 + I2)(R2 + R3 + R4) + I3(R3 + R4) + i_0 R3 + i_{0A2} R2$$
$$V_{GND} = I1R1 + (I1 + I2)(R2 + R3 + R4) + I3(R3 + R4) + i_0 R3 + i_{0A2} R2 + i_{0A1} R1$$

Figure 32.5   How incorrect grounding can lead to ground voltages other than 0 V

Since all three stages of the amplifier share the wire that provides the ground return for the power supply and the signal, voltages will be developed across the resistance that this wire presents to the currents. These will range from d.c. values for the power supply current for each stage to a.c. signals which are of varying phase. These will all add to make an error voltage at the ground connection of the stage, which can result in feedback effects and d.c. offsets, resulting in erroneous output signals, instability or distortion. The problem can be cured by reducing the resistance of the earth tracks to a very low level; in high-frequency circuits this is often seen as a continuous ground plane on one or other side of a PCB or sometimes as a buried layer. The effect is even more pronounced at very high frequencies when the circuit operation can be improved by the addition of individual stage screening. Figure 32.6 shows another useful approach called star point grounding. This is particularly appropriate where an output stage may have significantly higher current drain than other stages, and is often used in audio amplifiers, etc.

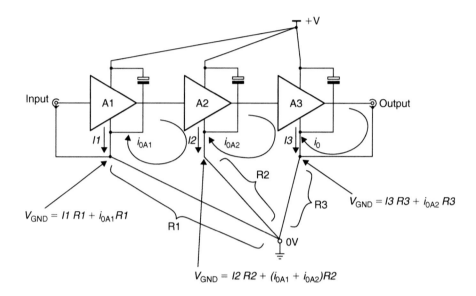

Figure 32.6   Improved circuit performance through the use of star topology for ground and supply

Another grounding effect is earth loops between equipment. This may be seen on the test bench if an oscilloscope and a power supply without an isolated output are used to test an amplifier circuit. The power supply and the oscilloscope both provide paths to mains earth from the circuit under test, and this loop can result in quite large signals being impressed on the circuit under test, resulting in unexpected or apparently random effects. The use of a double-insulated mains power supply is common in audio equipment as it avoids the hum resulting from loops returning to mains earth.

CE marking and the Low-Voltage Directive

The UK is bound by regulations developed in the EU; these include the EMC Directive and the Low-Voltage Directive (**LVD**) and various other directives like the Radio and Telecommunications Terminal Equipment (**RTTE**), Restrictions of Substances Hazardous to Health (**RoHS**) and Waste Electrical and Electronic Equipment (**WEEE**). Each of these directives is a very complex document and their implications are closely allied in their relationship to the electronics and electrical manufacturing and servicing industries. The CE marking signifies that the product or equipment to which it refers meets all the requirements of all relevant EU harmonized directives, including the health and safety features, complies with the appropriate regulations of LVD EN60950 and meets the emissions and immunity of EN50081/2.

Even the relative dimensions of the two circles that form the basis of the CE marker characters are specified (Figure 32.7): the characters of the mark must be at least 5 mm high.

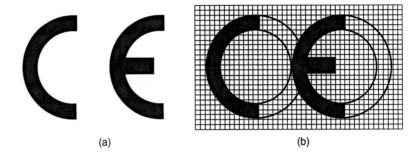

(a)   (b)

Figure 32.7 (a) CE mark applied to equipment indicating that it meets *all* the requirements of *all* relevant EU harmonized directives, and (b) the specification for the CE mark shape

> The standards have to be maintained throughout the product or equipment's lifetime, and also apply to second users and repaired items. This is why the requirements have to be observed in the course of servicing.

The **LVD** applies to items rated at between 50 and 1000 V a.c., and 75 and 1500 V d.c. It therefore applies to all domestic entertainment equipment and most of the industrial equipment so powered.

The EN60950 directive, which applies to information technology (IT) systems and any equipment that may be connected to an IT system, is a lengthy document that chiefly relates to matters of safety, including flammability, for users, operators and service personnel. In the interests of brevity, only the most important features of the directive can be included here. For specific information the reader is referred to Further reading list at the end of this chapter.

All equipment is assumed to be in continuous operation unless otherwise stated on the rating label. Virtually all IT equipment is covered by this legislation, including class 1, 2 and 3 devices with pluggable connectors, plus fixed, portable and handheld equipment.

The maximum earth leakage current is 3.5 mA, which should be the trip setting of any associated earth leakage circuit breakers (**ELCBs**). Even lithium batteries commonly used for memory backup are specifically mentioned in the text, because these can be explosive if they are incorrectly replaced. During servicing, care must be exercised because equipment such as cathode-ray tube monitors, some laser printers and photocopiers can include extra high voltages that exceed the LVD levels.

The LVD ratings are not intended for use in the following cases, which are covered by other international specialist safety provisions:

- in ships, aircraft, railways, explosive atmospheres, radiological and medical applications
- in domestic plugs and sockets and electricity supply meters
- in electric fence controllers and parts for goods and passenger lifts.

In addition, the directive does not necessarily apply to electrical equipment intended for export to non-EU countries; although many countries have adopted the EU rules as the basis of their own legislation, some have even tighter regulatory regimes.

The CE marking is the only marking that may be used to certify that a manufactured product conforms to the standards incorporated in the relevant directives. It is a criminal offence not to mark a product that is covered by LVD or so mark a product that does not comply.

Equipment may be provided on the market only if, having been constructed in accordance with good engineering practice in the safety matters in force within the community, it does not endanger the safety of persons, domestic animals or property when properly installed, maintained and used in the applications for which it was intended.

Third party conformity assessment is normally carried out by a recognized test house. These organizations, which can prove their conformity with the EN45000 series standards, which cover the criteria for the operation and standards testing of new equipment, are designated by the member states. These bodies may submit an accreditation certificate together with other documentary evidence. Under this system, the manufacturer takes full responsibility for the assessment, testing, documentation and declaration of conformity and the CE marking. A technical file (see below) or documentation must also be available on demand for the national enforcement authorities.

The RTTE directive covers devices such as mobile phones, wireless and wired networking equipment, remote keyless entry devices that use radio signals, CB and PMR equipment, and domestic radio and television receivers. The directive sets up a harmonized European framework of frequency bands and power limits with the aim of allowing the free movement of goods while preventing interference to services.

The **RoHS** directive sets requirements on the chemical composition of the materials used in the manufacture of equipment, and focuses on the materials that are embodied in the equipment or discharged into the environment as a result of manufacture. This and the associated **WEEE** directive, which defines how equipment must be dealt with at the end of its useful life, are intended to control the normal processes and products of manufacturing.

To certify conformity with the directives and the CE marking, the manufacturer must draw up a technical construction file covering the design, manufacture and operation of the equipment, prepare an EC declaration of conformity and affix the CE marking.

Control of Substances Harmful to Health

Control of Substances Harmful to Health (**CoSHH**) deals with the safe use of chemicals and other materials, and accidents involving them.

The human effects of hazardous substances require specialist treatment and it is therefore important that first aiders should be chemically aware of the associated problems. Under the regulations, all harmful substances should be stored in a cool, dry place and used with safety concerns as the primary consideration. The person responsible for this store should also be a trained chemical first aider. The safety information provided by the manufacturer should be stored where it is readily accessible to the appropriate member of staff, who may need it in a hurry.

Common solvents such as trichloroethane, trichloroethylene and carbon tetrachloride (CTC) that have in the past been used for cleaning and degreasing are now banned under EU law. Because all chlorinated solvents have narcotic and anaesthetic properties, any such solvents that are still in use should be treated very carefully. (Chloroform, another member of the family, is a well-known anaesthetic.) High levels of exposure to this group of substances can lead to kidney and liver damage. Of the group, only trichloroethane is non-carcinogenic. Although these substances are non-flammable, when chlorinated fluids are exposed to heat, they liberate the chemical warfare gases known as phosgene and hydrogen chloride. The chlorinated solvents can enter the human body either by inhalation or by skin absorption.

The use of chlorofluorocarbons (CFCs) is progressively being phased out, largely because they are destroyers of the upper atmospheric ozone layer, which protects against ultraviolet radiation from the sun. Apart from refrigeration applications, CFCs have been used as degreasing agents and freezer sprays. The very low boiling point can create frostbite owing to rapid evaporation and can strip the fatty tissue from the surface of human skin. Therefore, great care needs to taken when using up old stocks of CFCs.

Hydrochlorofluorocarbons (HCFCs), which are being used as temporary replacements for CFCs, are very good and non-destructive cleaners, and are non-flammable. However, under extreme heat they break down to generate phosgene, hydrochloride and hydrogen fluoride gases, so need to be very carefully stored.

Ethyl alcohol or ethanol, and its relative methylated spirits, are very good cleaning agents but are highly flammable and highly toxic, and can cause liver damage if drunk. This group also absorbs water fairly easily and this slows down the evaporation process.

Isopropyl alcohol (IPA), which is highly flammable, is a relatively safe cleaning agent, although its near neighbour, *N*-propanol, is classified as toxic.

Petrol and benzene are both extremely good degreasing and cleaning agents, but are highly flammable and explosive. They are harmful to health via skin absorption and the respiratory system and are also carcinogenic.

White spirit, which is commonly associated with paints, is highly flammable and produces narcotic fumes, which are harmful if they enter the respiratory system.

Solvents such as xylene and toluene, which are constituents of some switch cleaners and adhesives, produce euphoria (the glue-sniffing syndrome) and lead to serious problems such as respiratory failure and death, damage to the central nervous system and liver damage through skin absorption. Acetone and amyl acetate are solvents that produce toxic fumes and are also a fire hazard.

Many of the adhesives and sealants used in the servicing environment are also likely to create health risks when used at elevated temperatures. Long periods of exposure to the fumes from epoxy resins (two-part mixes) can also be a source of liver problems. While silver loaded epoxy glues have a low resistance that makes them useful for refixing metal shields within a piece of electronic equipment, the period of exposure can be reduced by raising the temperature to about 40–50°C to cut the curing time in half.

Solder fluxes (rosin or colophony) produce fumes when heated that can cause industrial asthma. Since the low-fume alternatives are also suspect, it is most important that all soldering operations are as far as possible carried out under a fume extractor hood. Fume extractor kits that mount directly onto a soldering iron are available, but tend to make the work more difficult. Smoke extractor hoods incorporate a charcoal filter that needs to be replaced periodically. The fumes from the lead in solder are also problematical, but while a smoke hood is helpful, the use of lead is included in regulations that are not within the scope of COSHH.

As with all accidents, events that are covered by COSHH must also be recorded in the workshop accident books, which must be kept up to date by a responsible person and be available for inspection by the appropriate authority.

The EU directives referred to in this chapter cover several weighty volumes and are written in legalistic terms. The information presented above can only represent a brief summary of the salient points in each document. For further and more explicit explanations of the directives, the reader can usefully refer to the publications listed below.

Further reading

Kervill, G. Practical Guide to the Low Voltage Directive. Newnes, Butterworth-Heinemann, Oxford, 2000.

Lewis, G. Communications Technology Handbook, 2nd edn. Focal Press, Butterworth-Heinemann, Oxford, 1997.

Tricker, R. CE Conformity Marking. Butterworth-Heinemann, Oxford, 2000.

Williams, T. EMC for Product Designers, 3rd edn. Newnes, Butterworth-Heinemann, Oxford, 2001.

Multiple-choice revision questions

32.1 What defines a customer's view of product quality ?
  (a) quality control testing
  (b) design
  (c) fitness for purpose
  (d) cost.

32.2 When does 'infant mortality' affect products?
  (a) at the beginning of service life
  (b) during manufacture
  (c) after burn-in
  (d) when they are of low quality.

32.3 A surface-mounted electrolytic capacitor marked 5E6 has a resistor marked W5 connected in parallel with it. What are their values?
  (a) $56\,\mu F$ 25 V and $68\,k\Omega$
  (b) $56\,\mu F$ 16 V and $68\,k\Omega$
  (c) $5.6\,\mu F$ 16 V and $680\,k\Omega$
  (d) $0.56\,\mu F$ 6.3 V and $680\,k\Omega$.

32.4 What should you use to solder a surface-mounted component to a PCB?
  (a) a low-power soldering iron
  (b) a temperature-controlled iron with enough power to do the job quickly
  (c) an acid flux
  (d) a method that does not heat the component.

32.5 What is significance of the CE mark on a piece of electrical equipment?
  (a) it was made in Europe
  (b) it meets the LVD and EMC directives
  (c) it complies with all relevant EU directives at the time of manufacture
  (d) it was designed in Europe.

Unit 3

Outcomes

1. Demonstrate an understanding of power supplies, to component level

2. Demonstrate an understanding of amplifiers, to component level and apply this knowledge safely in a practical situation

3. Demonstrate an understanding of oscillators, multivibrators and waveform generator circuits, to component level.

33 Direct current power supplies

Direct current (d.c.) power supplies are required by most electronic systems, either powered from the alternating current (a.c.) mains supply or operating from batteries. The purpose of the power supply circuit is to provide a constant voltage or in some cases current with minimum ripple and other disturbances to the system. A supply may have to respond quickly to changes in the current demanded by the circuit without allowing the supply voltage to exceed the specified operating limits. The majority of d.c. power supply circuits are ultimately supplied from the mains a.c. supply, so one of the most important functions of the supply is to convert the a.c. into d.c., usually using diode rectifier circuits.

In all rectifier circuits (Figure 33.1) each diode will be reverse biased for half of the a.c. cycle and conducting for the other half. The amount of this reverse bias depends on the type of circuit used, and is greatest for a half-wave rectifier feeding into a reservoir capacitor.

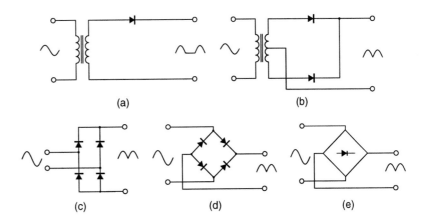

Figure 33.1 Rectifier circuits: (a) half-wave, (b) full-wave centre tapped, (c) bridge rectifier, (d) alternative drawing, and (e) symbol for bridge rectifier

Table 33.1 shows the operating results that may be expected from rectifiers of the three types described, half-wave, full-wave and bridge, given an a.c. input of E V peak ($0.7E$ V r.m.s.) and smoothing by a reservoir capacitor of adequate size. Note the large difference between no-load output voltage and the fully loaded voltage.

Table 33.1	Rectifier circuits compared for peak input E V			
Circuit full load	*d.c. output, no load*	*Max. reverse voltage, each diode*	*Ratio $I_{d.c.}/I_{a.c.}$*	*d.c. output,*
Half-wave	E	$2E$	0.43	$0.32E$
Full-wave	E	$2E$	0.87	$0.64E$
Bridge	E	E	0.61	$0.64E$

Note that in all rectifier circuits, reversing the connections of the diodes reverses the polarity of the output voltage.

Diode current and reservoir capacitors

Figure 33.2(a) shows a bridge rectifier with a capacitor connected across its output. The current flowing into and out of the capacitor (Figure 33.2b) is determined by the size of the load and the size of the capacitor. Note that it is very important not to use a capacitor that is too big because a very large capacitor draws a very large ripple current; this is why packaged diode bridge circuits usually specify the maximum capacitance in the data sheet.

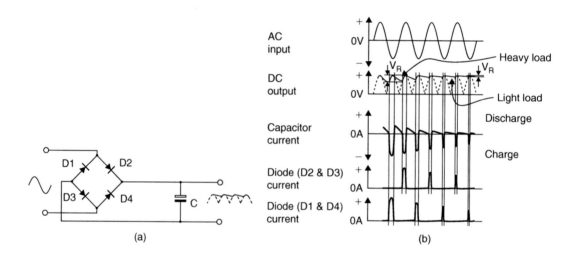

Figure 33.2 (a) Bridge rectifier with capacitor, and (b) waveforms at various loads

The simplest power supply circuits consist only of rectifiers and smoothing capacitors. In these circuits, the reservoir capacitor supplies the current to the load during the time when the diodes are cut off, and any ripple on the supply is reduced by filtering. Such circuits are adequate for many purposes, but they are too poorly regulated for use in circuits intended for measurement, computing, broadcasting or process control.

The **regulation** of a circuit is the term used to express the change of output voltage caused either by a change in the a.c. supply voltage or by a change in the output load current. A well-regulated supply will have an output voltage whose value is almost constant; the output voltage of a poorly regulated supply will change considerably when either the a.c. input voltage or the output current changes. Table 33.1 shows how the voltage of a smoothed supply changes from the no-load to maximum-load condition.

The change in output voltage caused by changes in the a.c. supply voltage is usually less great. A 10% change in the a.c. supply will also change the output voltage of a simple power supply by about 10%, the two percentage changes being almost identical.

Example: What will be the change in an unregulated 10V supply when the a.c. supply voltage changes from 240V to 220V?

Solution: The a.c. voltage change is 20V down on 240. The percentage change is therefore 20/240 × 100 = 8.3%. Since 8.3% of 10V is 0.83V, the 10V supply will drop to 10 0 – 0.83 V = 9.17V.

The effect of changes in the load current (which are themselves caused by changes in the load resistance) is more complicated, because there are two main causes (Figure 33.3). The first is the resistances of the rectifiers, the transformer windings and any inductors that may be used in filter circuits, and the internal resistance of the power supply. By Ohm's law, a change in current flow through such resistances must cause a drop in the output voltage.

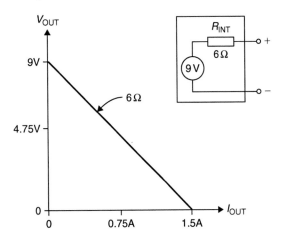

Figure 33.3 Load line of a supply: internal resistance of the supply causes the output voltage drop as the load current is increased

The second effect is the voltage drop that takes place across the reservoir capacitor as more current is drawn from it. This voltage drop (V) can be quantified approximately by using the equation:

$$V = \frac{It}{C}$$

where I is the load current (amps), t is the time (seconds) elapsing between one charge from the reservoir and the next, and C is the capacitance of the reservoir (farads).

Example: By how much will the voltage across a 220 μF capacitor drop when 0.2 A is drawn from a full-wave rectifying circuit?

Solution: In a full-wave rectifying circuit, the time between peaks is 10 ms. Substituting the data in the equation gives:

$$V = \frac{0.2 \times 0.01}{220 \times 10^{-6}} = 9\,\text{V}$$

A 50 V supply, for example, would give an output which drops to 41 V between peaks, and a 9 V peak-to-peak ripple at 100 Hz would be present.

Regulator circuits

A regulator circuit connected to a rectifier/reservoir unit ensures that the output voltage is steady for all designed values of load current or of a.c. supply voltage. Such a circuit can only do its job, however, if the rectifier/reservoir unit is capable of supplying the required output voltage (measured from the minimum of the ripple wave) under the worst possible conditions, when a.c. supply voltage is minimum and load current is maximum. The regulator will then prevent the output voltage from rising above this set value even when load current is small or the a.c. supply voltage is high.

A regulator requires a reference voltage that remains steady, and the action of the regulator is to compare the output voltage to this reference voltage and adjust the output accordingly. The reference voltage is obtained from a Zener diode.

Zener diode regulation

A **Zener diode** is used reverse biased so that the junction breaks down to permit current flow through the diode. This breakdown occurs at a precise voltage whose value depends on the construction of the diode, and causes no damage provided the current flow is not excessive. To prevent this, current must be regulated by connecting a resistor in series with the diode. The output of a Zener diode circuit is therefore a regulated voltage.

A simple regulator circuit is shown in Figure 33.4, with the output voltage across the diode being used to supply any other circuit or part of a circuit that requires a stable voltage. Many circuits, especially measuring and oscillator circuits, are adversely affected by voltage variation, which can be caused by changes in the supply voltage or changes in the current drawn by the load.

Open-circuit diode failures have the following effects:

- The d.c. output of rectifier circuits either is reduced or falls to zero. A half-wave circuit will have no output; full-wave and bridge circuits

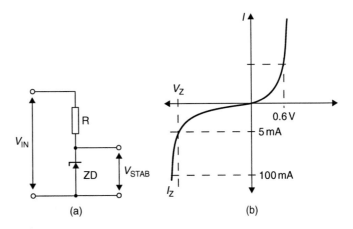

Figure 33.4 (a) Simple Zener diode shunt regulator, and (b) the Zener diode *I/V* characteristic curve

will have reduced output, with ripple at supply frequency if only one diode fails.

- A Zener diode regulator will give a higher voltage output, without regulation.

Short-circuit failures of diodes have the following effects:

- Rectifier circuits will blow fuses, and electrolytic capacitors may be damaged. A short-circuit diode will usually fuse itself, so becoming open-circuit. In a bridge circuit, a short-circuit diode will usually cause another diode to become open-circuit.
- A Zener diode regulator will give zero output.

Two basic types of regulator are used, a series circuit and a shunt circuit. The simplest example of the shunt type is a Zener diode regulator, illustrated earlier. In this type of circuit, the amount of current drawn from the power supply is constant. When load current is maximum, the regulator circuit current is minimum. When the load takes its minimum current, the regulator circuit takes its maximum current. The arrangement is therefore such that load current plus regulator current is always a constant value. The constant current flowing through R produces a constant voltage drop across it. So, if the input voltage remains unchanged, the output voltage will also remain constant.

The maximum dissipation of the Zener diode in this circuit is given by:

$$\text{Zener voltage} \times \text{Maximum current}$$

with dissipation in milliwatts if the current is measured in milliamps.

The maximum dissipation of the resistor is given by:

$$(\text{Unregulated voltage} - \text{Zener voltage}) \times \text{Maximum current}$$

Example: A 5.6 V (5V6) Zener diode is used to supply a load which takes a maximum current of 15 mA. If the minimum desirable Zener current is 2 mA, and the unregulated voltage is 12 V, find (a) the value of series resistance that must be used, (b) the maximum Zener dissipation, and (c) the maximum dissipation in the resistor.

Solution: With 15 mA flowing through the load and 2 mA through the Zener diode, total current is 17 mA. The voltage across the resistor is $12 - 5.6 = 6.4$ V, so that (a) the required resistance value, using Ohm's law, is $6.4/17 = 0.376$ K or 376 ohms. In practice, a 330 ohm resistor would be used, making the total current $6.4/.330$ mA $= 19.4$ mA. (b) At a current of 19.4 mA, the dissipation in the Zener diode is $5.6 \times 19.4 = 108.6$ mW, and (c) the dissipation in the resistor is $6.4 \times 19.4 = 124.2$ mW.

The shunt regulator is limited in its use by being inefficient and not good at regulating over a wide range of output currents. Usually, a better way of regulating the output of a power supply is by means of a series regulator in which a transistor is connected between the supply and the load (Figure 33.5). This is a much more common type of regulator circuit.

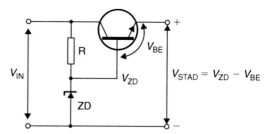

Figure 33.5 Basic series regulator circuit, sometimes called an amplified Zener regulator

When this type of regulator circuit is used, the transistor takes only as much current as the load, unlike the shunt regulator which takes its maximum current just as the load takes minimum current. The base voltage of the transistor is held constant by the regulator action of ZD_1 and R_1. The emitter voltage, and so the conduction, of transistor depends on the load voltage. If the demand for load current increases, the output voltage will tend to fall, increasing the forward bias. This allows it to pass more current to meet the demand. If the requirement for load current falls, this effect will be reversed.

Practical 33.1

Make up a simple series regulator circuit with the following component values: R = 330R, ZD = 5V6 and a 2N3055 transistor. V_{in} = 10 V. Connect this circuit to the power supply used previously, and draw a set of regulation curves for the stabilized circuit. Measure also the ripple voltage at maximum load current: (a) across the reservoir capacitor, and (b) across the load.

The power supply in Figure 33.6 includes output overcurrent protection with the conventional series regulator circuit, this protects the regulator from possible overheating and the powered equipment from the effects of overcurrent under fault conditions. In operation the output current flows through the 2 Ω resistor in series with the 0.5 A fuse. If the output current reaches about 0.3 A the voltage drop across the 2 Ω resistor begins to turn on the BC109 transistor, which 'steals' base current from the series pass transistor, reducing the output voltage and therefore the output current. This is usually described as soft limiting or foldback limiting.

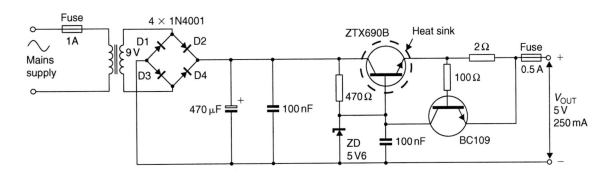

Figure 33.6 Mains power supply with output current limiter and Zener diode-based series regulator

More elaborate series regulator circuits use integrated circuit (IC) comparator amplifiers to drive the series transistor, and the whole circuit is in IC form. The use of IC regulators has now almost totally replaced discrete regulator circuits.

Integrated circuit regulators

Integrated circuit regulators are available in a wide variety of types, each tailored to specific circuit requirements, but the simplest are the familiar fixed voltage types such as the 7805. The final two digits of this type number indicate the stabilized output, 5 V for this example. The 7805 is a three-pin regulator which requires a minimum voltage input of 7.5 V to sustain regulation, with an absolute maximum input voltage of 35 V. The maximum load current is 1 A and the regulation against input changes is typically 3–7 mV for a variation of input between 7.1 and 25 V. The regulation against load changes is in the order of 10 mV for a change between 5 mA and 1.5 A load current. The noise voltage in the band from 10 Hz to 100 kHz is 40–50 µV, and the ripple rejection is around 70 dB. Maximum junction temperature is 125°C, and the thermal resistance from junction to case is 5°C/W.

The 78xx type of regulator is used extensively for power supplies in both linear and digital equipment A typical recommended circuit is shown in Figure 33.6, along with the diode bridge and reservoir capacitor. The capacitors that are shown connected each side of the IC are very important for suppressing oscillations and must not be omitted. In particular, the 100 nF capacitor at the input must be wired across the shortest possible path at the pins of the IC. Some types of regulator IC should not be operated without a load, since this can cause the circuit to oscillate at a high frequency. Often, a light-emitting diode (LED) and series resistor connected at the output to indicate that power is on will be sufficient to prevent oscillation.

The maximum allowable dissipation is 20 W, assuming an infinite heat sink, and the actual dissipation capabilities are determined by the amount of heat-sinking that is used. If no heat sink is used, the thermal resistance of the 7805 is about 50°C/W, and for a maximum junction temperature of 150°C this gives an absolute limit of about 2.5 W, which would allow only 2.5 V across the IC at full rated current, an amount only just above the absolute minimum voltage drop.

The thermal resistance, junction to case, is 4°C/W, and for most purposes, the IC would be mounted on to a 4°C/W heat sink, making the total thermal resistance 8°C/W. This would permit a dissipation of about 15.6 W, which allows up to 15 V or so to be across the IC at the rated 1 A current. This is very useful, since a 5 V supply will generally be provided from a 9 V transformer winding whose peak voltage is 12.6 V, making it impossible to cause overdissipation at the rated 1 A current, since the voltage output from the reservoir capacitor will be well below 12.6 V when 1 A is being drawn. It is important to be careful when specifying a transformer not to overspecify the output current, because the load line of a correctly rated transformer can be used to advantage in limiting the regulator dissipation.

The 78 series of regulators are complemented by the 79 series, which are intended for stabilization of negative voltages. The circuits that can be used are identical apart from the polarity of diodes and electrolytic capacitors, and the range of currents is substantially the same as for the 78 series.

If the regulated current that is required is greater than can be supplied from any available IC regulator, then the main options are either to use a regulator to control one or more power transistors rated to pass the required current, or to use a switched mode power supply (SMPS) circuit (see later), rather than a linear regulator.

In the early days of IC regulators, the failure rate of regulator chips was very high, but later versions have added **protective circuitry**, which has almost eliminated failures of the type that caused so many problems at one time. The main protective measures are for thermal protection and fold-back overload protection, and the effect in each case is that the output from the regulator will drop to a level that reduces risk to the regulator or to the equipment fed by it.

The **dropout** of a regulator is the minimum voltage difference that must exist between input and output to sustain the action of the regulator. The dropout for the 78 series of regulator ICs is at least 2 V, and for some purposes, particularly for stabilizing supplies that are based on secondary batteries, this differential is too large. Low dropout regulators allow much lower levels of input voltage, typically down to 5.79 V for a 5 V output regulator.

Low-dropout regulators were originally developed for the car industry to provide stabilized outputs for microprocessor circuitry, and they would not be used for general-purpose stabilization for mains-powered supplies. Most low-dropout regulators feature additional protection against supply reversal, the effects of using jumper leads between batteries, and large voltage transients. Several types also feature inhibit pins, which allow the regulator to be switched back on again after it has been switched off by an overload.

Practical 33.2

Construct an IC regulator circuit, as illustrated in Figure 33.7, using the 7805. Measure the output for input voltage levels that range from 6 to 12 V, and with load from 10 ohms to 1 K.

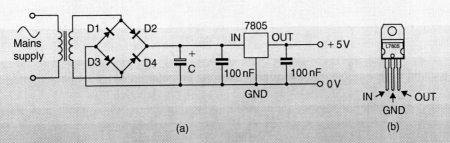

(a)   (b)

Figure 33.7   (a) Power supply circuit using the 7805 regulator, and (b) pin connections of the 7805 regulator IC

Failure in discrete regulator circuits is usually caused by excessive dissipation in the main transistor, shunt or series. In the series circuit the overdissipation will have been caused by an excessive load current, unless the regulator is protected against short-circuits.

An open-circuit Zener diode will cause a shunt regulator to cease conducting. Its effect on the series circuit is the opposite, in that the output voltage will rise to the level of the unregulated supply. A short-circuit Zener diode will cause the shunt circuit to pass excessive current. It will cause the series circuit to cut off.

Note carefully that all power transistors used in regulator circuits must be bolted to heat sinks of adequate size. Failure of IC regulators is usually due to thermal overload of regulators that do not have protection circuitry.

There are a few circuits that call for a high-voltage, low-current supply in which poor regulation is acceptable. Rather than wind a transformer especially to provide the high voltage, a **voltage multiplier** circuit (of which the voltage doubler is the simplest example) is often used instead. On the negative-going half of the voltage cycle shown in Figure 33.8 (left), C_1 is charged by current through D_1, so that point X is at a d.c. voltage equal to the peak voltage of the a.c. wave.

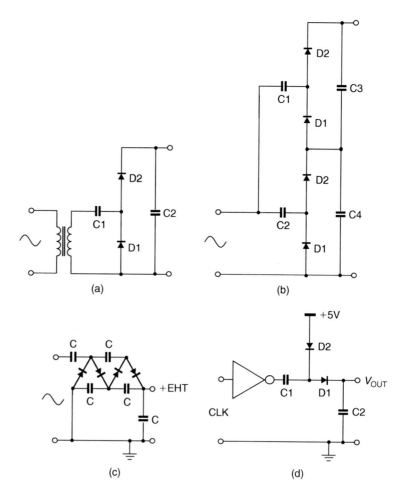

Figure 33.8 (a) Simple voltage doubler, (b) quadrupler, (c) Cockcroft–Walton multiplier, and (d) CMOS charge pump circuit

At the peak of the positive-going half-cycle, the peak inverse voltage across D_1 is equal to twice the peak voltage (the previous peak charge, plus the peak a.c. value). This causes C_2 to charge through D_2 to the same level. At line (supply) frequencies, the capacitor C_1 must be of large value, and must be rated at the full d.c. voltage. At higher frequencies, smaller values of capacitance can be used.

Multiplier circuits of this type are commonly used to supply the high voltages required by the colour tubes in colour television receivers. The circuit shown in Figure 33.8(b) is effectively two voltage doublers connected in series, while Figure 33.8(c) shows the Cockcroft–Walton multiplier circuit, named after its inventors. Circuits such as these are commonly used to obtain voltages as high as 25 kV for the final anode electrode of a colour cathode-ray tube. The complementary metal-oxide semiconductor (CMOS) gate-driven diode charge pump in Figure 33.8(d) operates on the same principles. This circuit is often used in battery equipment where a low current voltage outside the normal rails is required: think about how it can be made to generate a negative voltage from the positive rail.

Power supply unit components

The components used in power supplies must be adequately rated for the voltage and current levels that will be used. Electrolytic capacitors can be used for low-voltage smoothing applications, but they should not be used for high-voltage supplies, particularly above 500 V. For such applications, plastic dielectric capacitors must be used and for some specialized applications, oil-filled paper capacitors. Resistors are subject to a maximum voltage rating, often 350 V or less, and where higher voltages are used, chains of resistors connected in series must be used to keep the voltage drop across each individual resistor within the limit.

Power supplies that operate at low voltages usually deliver large currents, and computer power supplies in particular will supply 20 A or more at +5 V. Electrolytic capacitors used for smoothing such supplies must be rated to have low equivalent series resistance and be able to withstand high ripple currents. Some supplies make use of a series resistor for monitoring current, and this resistor will have a very low resistance value and is typically a special purpose type.

The use of switch mode circuits allows plastic dielectric capacitors to be used in smoothing circuits rather than electrolytics, because the ripple frequency is much higher.

Switch mode power supplies

The common configuration of linear-regulated power supply consists of a mains frequency transformer and rectifier, together with an IC series regulator. The latter simply behaves as a controlled series resistance to stabilize the output voltage. Such systems suffer from several serious disadvantages:

• They are most inefficient. It is unusual to find that more than 35% of the input energy reaches the load. The remainder is dissipated as heat. The inefficiency is greater for low-voltage, high-current supplies.

• The mains transformer is invariably large. Its size tends to be inversely proportional to the operating frequency.

• The reservoir and smoothing capacitors need to be large to keep the ripple amplitude within acceptable bounds. This is particularly difficult for low-voltage supplies.

- Because the series transistor (or transistors) is operated in the linear mode it must be mounted on a large heat sink.

If the operating frequency can be increased significantly, both the transformer and the filter capacitors can be reduced in size. If the series transistor can be operated either cut-off or saturated, its dissipation will be greatly reduced. The power supply can then be made more efficient. Such operation can be achieved using an SMPS. These circuits can operate with efficiencies as high as 85%.

The basic switching principle of the most common type of SMPS (sometimes called a **Buck converter**) is shown in Figure 33.9(a). When the switch is closed, current flows through the inductor or choke L to power the load and charge the capacitor C. When the switch is opened, the magnetic field that has been built up around L now collapses and induces an electromotive force (emf) into itself to keep the current flowing, but now through the flywheel or freewheel diode, D. The voltage across C now starts to fall as the load continues to draw current. If the switch is closed again the capacitor recharges. This switching cycle produces a high-frequency supply voltage.

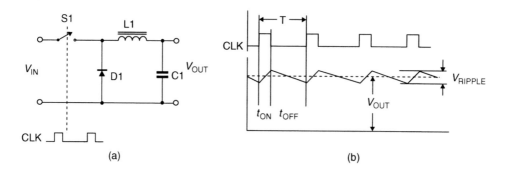

Figure 33.9   (a) Basic principle of an SMPS, and (b) switching waveforms

The duty cycle or switching sequence is shown in Figure 33.9(b) together with the output voltage V_{out}, that it produces. Increasing the on-period will increase V_{out}, whose average level is given by $V_{in} \times t_{on}/T$. V_{out} can be regulated by varying the mark-to-space ratio of the switching period, with any ripple being removed by a low-pass filter in the usual way.

The typical SMPS whose block diagram is shown in Figure 33.10 consists of a mains rectifier with simple smoothing whose d.c. output is chopped or switched at a high frequency, using a transistor as the switch. For television applications this switch is commonly driven at the line frequency of 15.625 kHz. The circuit generally needs some startup arrangement that will ensure drive to the PWM switch when no d.c. output exists.

For industrial applications or computer power supplies the switching frequency is usually in the order of 20–25 kHz. The chopped waveform is applied to the primary circuit of a high-frequency transformer that uses a

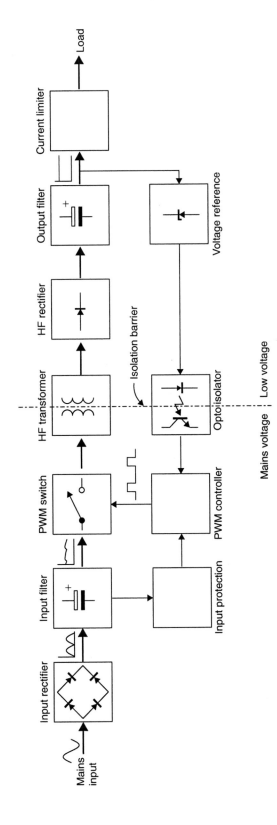

Figure 33.10   Simplified block diagram of switched mode power supply typical of the type that powers a PC

ferrite core for high efficiency. The signal voltage at the secondary is recti-fied and filtered to give the required d.c. output. This output is sensed by a control section that compares it with a reference voltage to produce a cor-rection signal, which is used in turn to change the mark-to-space ratio of the switching circuit to compensate for any variation in output voltage. This action is effectively pulse-width modulation.

The ripple frequency of 50 Hz at the input has been changed to a fre-quency of 20 kHz at the output so that the smoothing and filter capacitors can be reduced in value by the ratio 20 000:50, equal to 400 times. No elec-trolytic capacitors need to be used.

The oscillator/rectifier part of the circuit can be operated from a battery or any other d.c. input, so that it becomes a device for converting d.c. from a high voltage to one at a lower level. Another option is to use a transformer whose input is the chopped high-frequency voltage, with several outputs that are rectified to produce d.c. at different voltage levels. Only one of these levels can be sampled to provide control, so that only one output is stabilized against load fluctuations, although all are stabilized against input fluctuations.

Switched mode power supply circuits are universal in small computers, because of the need to regulate a low-voltage supply at a high current. The usual circuitry rectifies the mains voltage directly (using no input trans-former) so that the early stages operate at high voltage and low current, and a conventionally regulated supply is used to operate the control stages, ensuring that these are working at startup. A transformer for the high-frequency voltage provides for isolation from the mains and for voltage output of +5 V (main output at high current), along with –5 V, +12 V and –12 V. A complete SMPS circuit can be obtained in IC form, and for higher outputs an IC can be used to control a high-power switching transistor.

The SMPS generates more radiated and line conducted noise than a lin-ear supply. This can be reduced to acceptable levels by using:

- mains input filters balanced to earth to give rejection of the switching frequency
- suitable design of output filter
- electrostatic screen between primary and secondary of the mains transformer
- efficient screening of the complete unit.

Excess-voltage trip

Figure 33.11 shows the layout of a typical excess voltage protection circuit using the **crowbar** principle. The circuit consists of three components: a Zener diode, in this case a 68 V part, a resistor and a thyristor. When the output voltage is held at its normal level of, say, 48 V, the diode is reverse biased and non-conducting. Therefore, no voltage develops across the resis-tor, and the gate voltage of the thyristor is close to zero.

If the output voltage rises above 48 V, the Zener conducts, drawing current through the 220 Ω resistor, 2.8 mA giving a 0.6 V drop, and the gate voltage of the thyristor rises, triggering it into conduction. This applies a short-circuit

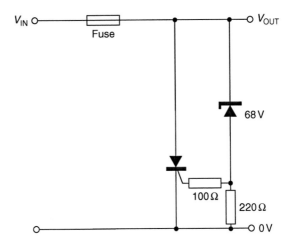

Figure 33.11 Crowbar: excess voltage trip circuit can be used to protect the load of a SMPS

across the supply input, which can either blow the mains input fuse or trigger a thermal cut-out to break the circuit. The effect of causing the conduction of the thyristor is therefore like placing a very low resistance (the crowbar) across the supply rails when an overvoltage condition occurs. Crowbar circuits can sense the output or input voltage of a supply, but the thyristor is usually placed at the input of the supply to blow the incoming fuse.

Overcurrent protection of SMPS can be provided both at the output using a sense resistor or at the controller by monitoring the transformer current waveform. Depending on the application and the speed of response required, both methods may be used.

Fault-finding and testing on any SMPS circuit can be difficult because of the way in which the units are connected to each other and depend on each other. This is particularly so when the SMPS is part of a television receiver for which the driving pulses for the SMPS are taken from the television circuits. One important rule is that you should not attempt to operate the SMPS without a load.

If you need to isolate the SMPS for testing, you should connect a dummy load, and the ideal type of dummy load is an incandescent lamp with the appropriate power and current rating. Another safeguard is to obtain the input from a Variac™ autotransformer, allowing the a.c. input to be increased gradually. Yet another way of protecting the SMPS while under test is to include a 60 W lamp in series with the mains supply. The positive temperature coefficient of the lamp will guard against surges and protect against short-circuits.

The behaviour of SMPS can be effectively studied using SMPS kits. These not only provide valuable construction exercises and are useful as transistor–transistor logic (TTL)-type power supplies, but also can be used to study the regulation and ripple characteristics of SMPS. In the absence of a suitable training kit, an ideal substitute is a power supply unit from a computer. These are now inexpensive and readily obtainable items that can provide considerable experience with SMPS, although it is not always easy to obtain a circuit diagram.

You should record carefully voltages and waveforms for the SMPS working normally, and also in various fault conditions.

Inverters and converters

Although **inverters** are primarily designed to convert d.c. power into a.c. power, they have some features in common with **converters**, which are used to transform d.c. energy from a low-voltage level into d.c. at a higher voltage or d.c. of reverse polarity. The block diagram of Figure 33.12 shows some of these common features. An oscillator, usually running at an ultrasonic frequency (typically around 20 kHz), is powered from a d.c. source. This oscillator drives a transformer either to provide an a.c. output in the case of an inverter, or to power a rectifier/smoothing circuit for

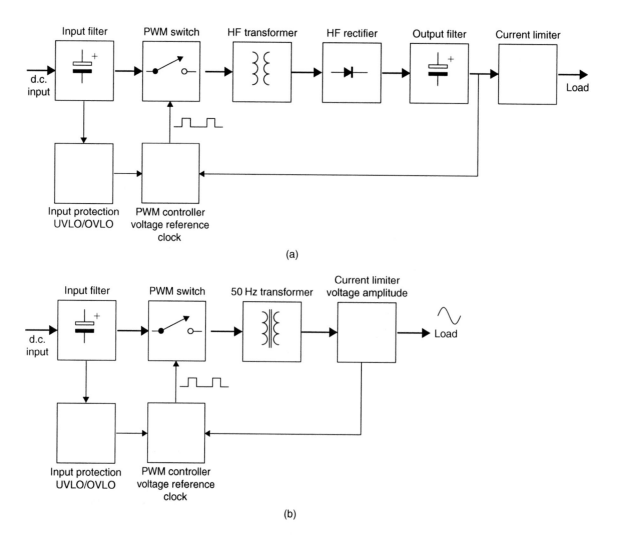

Figure 33.12   Block diagram of: (a) d.c.–d.c. converter, and (b) an inverter

converter applications. It is important to note that in each case the total output power must be lower than that supplied from the d.c. source owing to the energy losses in the transformer, etc. Using a rectifier bridge and smoothing or regulation in conjunction with an inverter produces a converter. Typical applications include producing d.c. at a higher or lower voltage or inverting polarity, so that a +5 V input can provide a –5 or –12 V output.

Many low-power, up to about 300 W, mains inverter designs are based on the use of an audio amplifier driving a mains transformer secondary with a 50 Hz sine wave so that the output from what would have been the mains primary is a 240 V r.m.s. sine wave at 50 Hz. Such systems typically use a full load drive current of 15–25 A r.m.s. from a 12 V source such as a car battery. Higher power inverters tend to be switch mode designs with a 50 Hz sine wave modulating the much higher frequency drive signal. This greatly reduces the power dissipation, as efficiencies approaching 85% can be achieved and much lighter and smaller ferrite transformers can be used.

Interference reduction

When using mains-powered equipment it is important that the minimum of noise and interference generated should be fed back into the mains. This is particularly important where data-processing equipment is operating in close proximity to high-current industrial plant. Care is also needed where linear circuits are operating at low signal levels. Power supplies of the SMPS or oscillator types in particular should have their inputs well filtered to prevent pollution of the mains supply. For some linear circuits, the frequency of operation of the SMPS may have to be carefully chosen to avoid interference.

Interference can be continuous (as for SMPS) or transient (as when switching a supply on or off), and both radio frequency and transient suppression will be needed for most types of power supply.

Typical filter circuits and component values are shown in Figure 33.13 (a–c). Inductors and capacitors used in these applications should have adequate working voltage and/or current ratings under fault conditions. The

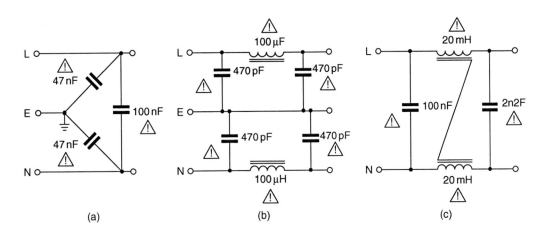

Figure 33.13 Typical line interference filters: (a) capacitor hf bypass, (b) differential low-pass filter, and (c) common-mode low-pass filter

characteristics of such filters are low pass, with zero attenuation at 50 Hz and at least 30 dB over the frequency range 150 kHz to 50 MHz. Data processing equipment is particularly susceptible to mains-borne interference. Data may be corrupted by the high voltage transients induced from inductive loads on the mains supplies. Fortunately, this problem is fairly easily solved by wiring devices called **varistors** across each pair of the mains supply wiring.

Varistors are particularly useful for transient suppression. These components are usually manufactured using zinc oxide or silicon carbide, with the former often being preferred because of its faster response. A varistor has a high resistance below some critical voltage, but above this, it rapidly conducts to short-circuit any large overvoltage condition. The working voltages of these devices, which can be used for both a.c. and d.c., range from about 60 V up to about 650 V.

Mains power outlets and mains lead plugs are now available for low-power (less than about 2 kW) applications. These contain both filters and surge suppressors. They are particularly suited to the mains supplies for minicomputer and microcomputer installations.

Power supply units and other circuits that operate at high dissipation will usually contain safety-critical components. These are typically resistors with specified dissipation characteristics or capacitors that are constructed to avoid dangerous breakdown, and such critical components are usually marked as such on circuit diagrams.

The important point about such components is that they must be replaced, if found to be faulty, with the identical type of part, never with an ordinary component that seems to offer similar characteristics. Since a breakdown of such a component compromises safe use of the equipment, anyone replacing such a component with an ordinary one could be held responsible for the consequences of equipment breakdown.

Multiple-choice revision questions

33.1 If a bridge rectifier is driven from an a.c. source, with peak voltage V'_{ac}, what is the peak voltage across each diode?
  (a) $2V'_{ac}$
  (b) $1.414V'_{ac}$
  (c) V'_{ac}
  (d) $0.5V'_{ac}$.

33.2 A mains power supply uses a 24 V transformer and bridge rectifier, with a 10 000 μF smoothing capacitor. If the output current is 1 A, what is the approximate output ripple voltage?
  (a) 1 Vpp
  (b) 0.7 Vpp
  (c) 0.31 Vpp
  (d) 0.1 Vpp.

33.3 In the previous question the mains frequency is 50 Hz. What is the ripple frequency?
  (a) 25 Hz
  (b) 50 Hz
  (c) 100 Hz
  (d) 200 Hz.

33.4 A shunt regulator circuit using a Zener diode is required to provide a load with a current of 5 mA at 6.8 V, when the input voltage can vary between 9 and 12 V (10 NiCd cells in series). What series resistor is required if the Zener requires a current of at least 5 mA for regulation?
  (a) 470 Ω
  (b) 220 Ω
  (c) 180 Ω
  (d) 160 Ω.

33.5 In question 4, what is the maximum power dissipated in the Zener diode?
  (a) 34 mW
  (b) 68 mW
  (c) 126 mW
  (d) 160 mW.

33.6 A series amplified Zener regulator supplies 5 V to a load at a current of 1 A, what is the dissipation in the pass transistor if the unregulated supply delivers 8.4 V at 1 A?
  (a) 1 W
  (b) 3.4 W
  (c) 5 W
  (d) 8.4 W.

34 Analogue amplifiers

The bipolar junction transistor (BJT) and the field-effect transistor (FET) both control the current flow at their output terminals (collector and drain, respectively), and in both cases this current at the output can be controlled by the voltage at the input.

The ratio (Change in current at output/Change in voltage at input), with a constant supply voltage, is called the **mutual conductance** of the particular bipolar transistor or FET to which it applies. Its symbol is g_m. The values of mutual conductance obtainable from bipolar transistors are much greater than are those from FETs. In the mutual conductance graph shown in Figure 34.1, for instance, a 30 mV input wave gives a 1 mA current flow at the output. The mutual conductance, g_m, is therefore $1/0.03 = 33.3$ mA/V. Metal oxide semiconductor field-effect transistor (MOSFET) circuits are used for small-signal amplifiers mainly in integrated circuit (IC) form, although power MOSFETs are found in hi-fi amplifiers. The examples here are mainly for BJT circuits.

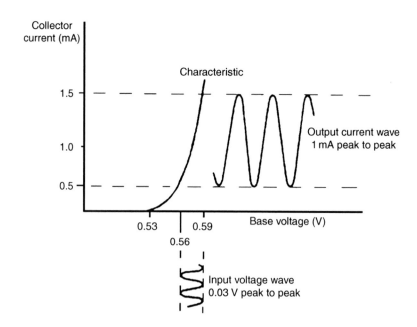

Figure 34.1 Mutual conductance graph with input and output waves indicated

The value of g_m for a BJT depends mainly on the steady value of collector current, I_c, and is approximately $40 \times I_c$ mA/V for any silicon transistor. If the collector load is R_c, then the gain is $40 \times I_c \times R_c$. Gain can be maximized by raising either the steady collector current or the collector load resistance, or both.

The method of operation for signal amplification is as follows. A signal voltage, alternating from one peak of voltage to the other, at the input produces a signal current, alternating from one peak of current to the other, at the output. To convert this signal current into a signal voltage again, a load is connected between the output terminal and the supply voltage. In a d.c. or an audio amplifier, a resistor can be used as the load, but for intermediate frequency (IF) or radio frequency (RF) amplifier circuits a tuned circuit (which behaves like a resistor at its tuned frequency) is used instead.

When the connection is in the common-emitter (CE) mode, the use of a load resistor causes the output voltage to be **inverted** compared to the input voltage. For example, a *higher* steady voltage at the input causes a greater current flow at the output. A larger voltage is therefore dropped across the load resistor, which causes the output steady voltage to be *lower*.

Any amplifying stage can give current gain, voltage gain and power gain, and the amounts of gain that can be obtained depend on the way in which the circuits are connected. Figure 34.2 shows the three possible amplifying connections of a single transistor, with their relative gain values (bias components not shown).

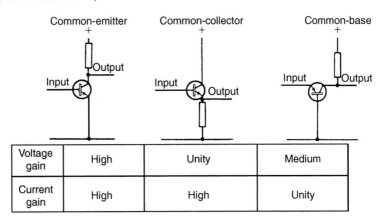

Figure 34.2 Amplifying connections of a single transistor

Note that circuits using active devices such as bipolar transistors or FETs give **power gain**. The amount of this gain is calculated by multiplying voltage gain by current gain. Passive devices such as transformers can provide voltage gain or current gain, but not power gain. The additional power is supplied by the d.c. supply to the transistor.

The behaviour of an amplifier can be clearly read from a graph of its output voltage or current plotted against its input voltage or current (for given values of load resistance and supply voltage). For example, the transfer characteristic of a small bipolar transistor is shown in Figure 34.3. An input current wave of $50\,\mu A$ peak-to-peak (p-p) produces an output current wave of $4\,mA$ p-p. The current gain h_{fe} of the transistor under these conditions is thus:

$$\frac{\text{Collector current swing}}{\text{Base current swing}} = \frac{4\ mA}{50\ \mu A} = 8$$

(Note that $4\,mA = 4000\,\mu A$.)

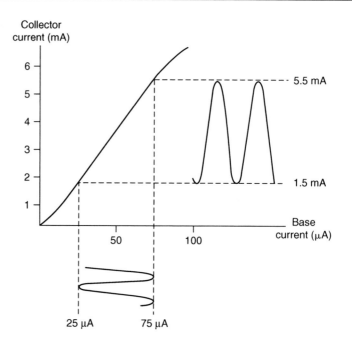

Figure 34.3   Transfer characteristics of a small bipolar transistor

The current wave that is produced is in this case an exact copy of the input wave, because the part of the graph that represents the values of input and output current is a straight line. Because a straight-line characteristic produces a perfect copy, such an amplifier is called a **linear amplifier**. If the input signal had been greater, with peak values 0 and 100 μA, the output signal would not have been a perfect copy of the input wave because these values of input would make use of the part of the characteristic that is not a section of a straight line. A transistor working in linear conditions is also said to be working in *class A* (see later).

Figure 34.4 shows the plots of the output current/input voltage characteristics of (a) a bipolar transistor and (b) a junction field-effect transistor

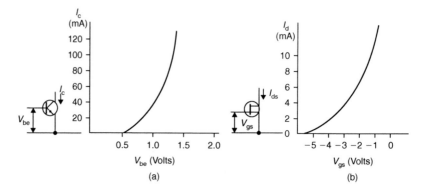

Figure 34.4   Mutual characteristics of (a) a typical bipolar transistor, and (b) a typical junction FET

(JFET). The curved shape of these characteristics shows that reasonably good linear amplification is possible only if a small part of the characteristic is selected for use. It is not, for example, possible to use an input voltage of less than 0.6 V for the bipolar transistor.

Another way of looking at transistor or FET action is by drawing the **output characteristic**. In the illustration of Figure 34.5 each line is the graph of I_c plotted against V_c for a given base current I_b. The difference in spacing shows that amplification will be non-linear. In other words, if the output characteristic lines are drawn for equal changes in input voltage or current, the unequal spacing of the lines indicates that the transfer characteristic must be curved, producing non-linear effects. Such effects produce **distortion**, meaning that the shape of the signal wave will be altered.

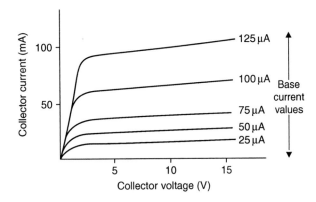

Figure 34.5 Typical output characteristic

Bias

The mutual characteristics of the bipolar transistor shown in Figure 34.4(a) made it clear that if such a transistor is to be used as a linear amplifier, the output current must never cut off, nor must the output voltage be allowed to reach zero (its bottomed condition).

Ideally, when no signal input is applied, output current should be exactly half-way between these two conditions. This can be achieved by **biasing**, by supplying a steady d.c. input which ensures a correct level of current flow at the output. A correctly biased amplifier will always deliver a larger undistorted signal than will an incorrectly biased one.

A correctly biased amplifier operating in such conditions, with the output current flowing for the whole of the input cycle, is said to be operating under **class A** conditions.

Practical 34.1

Connect up the circuit shown in Figure 34.6, in which Tr₁ can be any medium-current silicon NPN transistor such as 2N3053, 2N1711, 2N2219 or BFY60. Set the signal generator to deliver a signal of

(Continued)

Practical 34.1 (Continued)

50 mV p-p at 1 kHz when connected to the amplifier input. Connect the oscilloscope to the output terminals, with the Y-input of the oscilloscope to point X. Set the oscilloscope Y-input to 1 V/cm and the time-base control to 1 ms/cm, and switch on the oscilloscope. When the trace is visible, adjust the potentiometer V_{r1} to its minimum voltage position, and switch on the amplifier circuit.

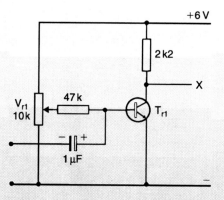

Figure 34.6   Suggested circuit for Practical 34.1

Note that there is no output from the amplifier because it is incorrectly biased. Gradually increase the bias voltage by adjusting V_{r1} until a waveform trace appears. Draw the waveshape. Continue to adjust the bias voltage until the waveform seen on the oscilloscope screen is a pure sine wave. It may be necessary to adjust the amplitude of the input wave to achieve this. Then increase the bias still further until distortion becomes noticeable again, and sketch this waveform also. You will see that with too little bias the amplifier cuts off, causing the top of the waveform to flatten. With too much bias, the amplifier bottoms, causing the bottom of the waveform to flatten.

Bias circuits

Three types of bias circuit are illustrated in Figure 34.7. The simplest uses a single resistor connected between the supply voltage and the base of the transistor (Figure 34.7a). This type of bias is not used for linear amplifier stages because it is difficult to find a suitable value of bias resistor and the bias is greatly affected by changes in transistor characteristics caused, for example, by changes in temperature.

The circuit shown in Figure 34.7(b) represents a considerable improvement, because the bias resistor is returned to the collector of the transistor rather than to the fixed voltage supply line. This small change makes the bias self-adjusting because of **d.c. negative feedback**, and the bias is now said to be stabilized.

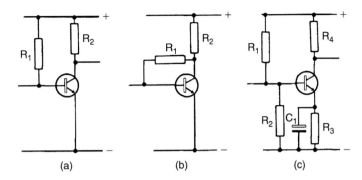

Figure 34.7   Bias systems: (a) simple, (b) current feedback, and (c) fixed base voltage type

The third bias circuit, shown in Figure 34.7(c), is the most commonly used of all for discrete transistor amplifier circuits, either BJT or FET. The negative feedback system of biasing is the main (and usually the only) method of biasing IC amplifiers (see later in this chapter). A pair of resistors is connected as a potential divider to set the voltage at the base terminal, and a resistor placed in series with the emitter controls the emitter current flow by d.c. negative feedback. Note that the emitter current is practically equal to the collector current flow ($I_c = I_e + I_b$, but I_b is very small).

In this type of circuit, the replacement of one transistor by another has little effect on the level of steady bias voltage at the collector, and changes caused by altering temperature similarly have very little effect. This biasing arrangement is therefore ideal for use in mass-produced circuits which must behave correctly even when fitted with transistors having a wide range of values of h_{fe}.

Figure 34.8 illustrates the bias method (a) which is used for a FET in the depletion mode, and the circuit (b) commonly used for MOSFET biasing.

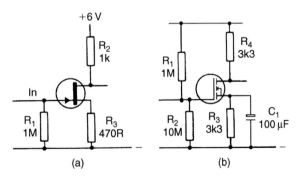

Figure 34.8   Biasing circuit for: (a) junction FET, and (b) MOSFET

For correct bias, the voltage of the gate for the FET types illustrated here should be negative with respect to the source voltage or, to put it another way, the source voltage must be positive with respect to the gate voltage. In this circuit the positive voltage is obtained from the voltage drop across the resistor R_3 in series with the source. The gate voltage of the JFET is kept at earth (ground level, or zero volts) by the resistor R_1, which needs to have a very large value since no current flows in the gate circuit. The MOSFET circuit uses a potential divider (R_1 and R_2) to maintain a steady voltage on the gate, and the source is biased by the current flowing in R_3. The calculations for the biasing of a FET are considerably simpler than for the biasing of a bipolar transistor.

Bias failure

Bias failure can be caused by either open-circuit or short-circuit bias components. In any of the circuits shown in Figure 34.7, if a short-circuit develops across the resistor R_1, the large bias current that will flow in consequence will cause the collector voltage to bottom, and may burn out the base-emitter junction. If the base-emitter junction thus becomes open-circuit, collector voltage will rise to supply voltage, so that the same fault can be the cause of either symptom. If R_1 becomes open-circuit, there is no bias supply and the collector voltage cuts off, so that collector voltage equals supply voltage.

In the case of the circuit in Figure 34.7(c), faults in resistors R_2 and R_3 can also affect the bias. Table 34.1 summarizes the possible faults and their effects.

Table 34.1   Bias faults and their effects

Fault	Collector voltage	Emitter voltage
R2 o/c	Low	High
R2 s/c	High	Zero
R3 o/c	High	High
R3 s/c	Low	Zero

o/c: open-circuit; s/c: short-circuit.

Gain and bandwidth

The voltage gain (G) of an amplifier is defined as:

$$G = \frac{\text{Signal voltage at output}}{\text{Signal voltage at input}}$$

with both signal voltages measured in the same way (either both r.m.s. or both p-p). This quantity G is an important measure of the efficiency of the amplifier and is often expressed in decibels by means of the equation: $dB = 20 \log G$.

Example: Find the gain of an amplifier in which a 30 mV p-p input signal produces a 2 V p-p output signal.

Solution: Insert the data in the equation $G =$ (Signal voltage at output/Signal voltage at input), so that $G = 2000/30 = 66.7$. Note that the 2 V must be converted into 2000 mV, so that both input and output signals are quoted in the same units. Expressing the same answer in decibels, $G = 20 \log 66.7 = 36.5 \, dB$.

A decrease in gain from 66.7 to 60 might seem significant, but the same decrease expressed in decibels is only from 36.5 dB to 35.5 dB, a change of 1 dB, which is the smallest change of gain that can be detected by the ear when the amplifier is in use. Measurements of gain expressed in decibels can therefore show whether changes of gain are significant or not. Figures of voltage gain by themselves are often misleading for this purpose.

One very useful figure is the **gain-bandwidth product**, which is the result of multiplying gain by the frequency bandwidth (between the 3 dB points; see the following section). This is often quoted for transistors to allow estimation of the maximum possible bandwidth or the gain at different bandwidths. No amount of circuit tricks can increase the basic gain-bandwidth product of a transistor, a point that must be kept in mind when one transistor is substituted for another.

Frequency response   Voltage amplifiers do not have the same value of gain at all signal frequencies. The circuit of Figure 34.9 shows the components in a single-stage transistor amplifier that determine frequency response. (C4 is shown dotted because it consists of stray capacitances and is not an actual physical component.)

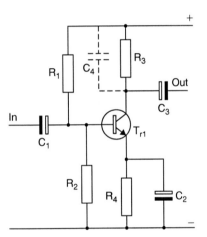

Figure 34.9   Components that affect frequency response

In the circuit, C_1 prevents d.c. from the signal source from affecting the bias at the base of the transistor, and C_3 prevents d.c. from its collector from affecting the next stage. The circuit can therefore provide no voltage gain for d.c., and the amount of gain it can give at low a.c. frequencies is inevitably limited by the action of the capacitors C_1 and C_3. Gain for low frequencies will also be affected by C_2, because this capacitor bypasses the negative feedback action of R_4 for a.c. signals only.

At the high end of the frequency scale, the stray capacitances that are present at the collector of the transistor and in any circuit connected through C_3 are represented by C_4 connected across the load resistor. These strays act to bypass high-frequency signals, so that gain decreases at these frequencies also. Only in the medium range of signal frequencies most commonly used is the gain given by this circuit configuration constant. This constant value is often referred to as **mid-range gain**.

Note that raising the collector load resistance will provide more gain (if the steady bias current is not reduced), but will decrease the bandwidth.

A typical curve of gain (in decibels) plotted against frequency for such an amplifier is shown in Figure 34.10. In this frequency–response graph, note that the frequency scale is logarithmic, so that 10-fold frequency steps occupy equal lengths of horizontal scale. This type of scale is necessary to show the full frequency range of an amplifier, and it normally plots gain in decibels.

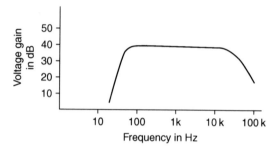

Figure 34.10   Typical curve of gain plotted against frequency, for an untuned amplifier

Note that these gain/frequency graphs apply only to sine-wave signals. The response of any amplifier to sudden steps in voltage (transients) is affected by the **slew rate** of the amplifier, a topic that will be dealt with later, under the heading of Operational amplifiers.

When a tuned circuit is used as the output load of an amplifier (Figure 34.11), the shape of the gain/frequency graph becomes more peaked. The

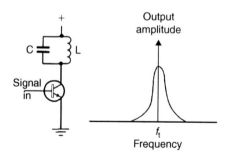

Figure 34.11   Using a tuned circuit as a load and the typical output/frequency graph shape

reason is that the tuned circuit presents a high resistance to the signal at the frequency of resonance (f_r). At resonance, this resistance has a value of L/CR ohms, where L is in henries, C in farads, and R is the resistance of the coil expressed in ohms. This value, L/CR, is called the **dynamic resistance** of the tuned circuit.

At all other frequencies, the load has a considerably lower value of resistance and acts instead like an impedance, with the result that voltage and current become out of phase with one another.

The gain of a tuned amplifier at the resonant (tuned) frequency is often controlled by an automatic gain control (AGC) voltage. This voltage is applied at the base (or gate of a MOSFET), either opposing or adding to the existing d.c. bias. **Reverse AGC** uses a d.c. bias voltage that reduces normal bias current, while forward AGC uses a bias voltage that increases the normal bias current. The type of AGC used depends on the type of transistor and the circuit of which it is a component.

To give **forward AGC**, a resistor is included in series with the load, but bypassed by a capacitor, so that signal current does not flow through it, and the transistor is so designed that its gain is much lower at low collector voltages. Increasing the bias current then lowers collector voltage and decreases the gain.

Most transistors, however, give greater gain at higher bias currents (given unchanged collector voltage), and so need to use reverse AGC.

Practical 34.2

Use a single-stage transistor amplifier (designed so that the upper 3 dB point it not too high, for easy measurement), along with a signal generator, attenuator and oscilloscope to measure the lower and upper 3 dB frequencies. Measure the gain at the middle of the range (i.e. halfway between the lower and the upper 3 dB points) and calculate the gain-bandwidth. Make sure that the input is low enough to avoid noticeable signal distortion at the output. After making your measurements, increase the signal input until the output becomes distorted (not a sine wave). Sketch the output wave shape.

For an audio amplifier, the lower and upper 3 dB points (f_1 and f_2) are quoted, so that such an amplifier can be described as being, for example, 3 dB down at 17 Hz and at 35 kHz. For a tuned amplifier we usually quote the tuned frequency and bandwidth; for example, a bandwidth of 10 kHz centred on 470 kHz.

Table 34.2 lists faults that have predictable effects, especially on gain and bandwidth.

Table 34.2 Amplifier faults	
Fault	*Effects*
Emitter bypass capacitor o/c	Reduced gain, increased bandwidth
Collector load resistor too low	Low gain, increased bandwidth, d.c. voltage too high
Underbiased transistor	Gain reduced, signal distortion at output

Multiple-stage amplification

For most amplifier applications a single transistor is not enough to provide sufficient gain, and several stages of amplification are needed. When an amplifier contains several stages, its total gain, G_t, is given by the equation $G_t = G_1 \times G_2 \times G_3$, where G_1, G_2 and G_3 are the gains of the individual stages.

In decibels, this becomes: Total gain = $dB_1 + dB_2 + dB_3$, giving the total gain in decibels. Note that the decibel figures of gain are *added*, whereas the voltage (or current, or power) figures have to be *multiplied*. This is because logarithms are used in the construction of the decibel figure, and addition of logarithms is equivalent to multiplication of ordinary figures.

The coupling together of separate amplifying stages involves transferring the output signal from one stage to the input of the next stage. This can be done in several ways, as described below and illustrated in Figure 34.12 (from which details of all biasing arrangements have been omitted for the sake of clarity). The aim is to transfer the maximum amount of signal from one stage to the next.

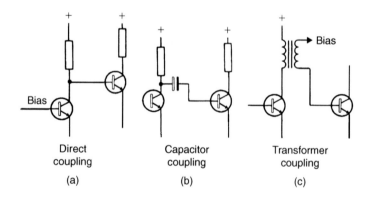

Figure 34.12   Coupling amplifying stages

Direct coupling involves connecting the output of one transistor to the input of the next, using only resistors or other components that will pass d.c. The result is that both d.c. and a.c. signals will be coupled. A d.c.-coupled amplifier by definition amplifies d.c. signals, so that a small change in the steady base voltage of the first stage will cause a large change in the steady collector voltage of the next. In all d.c.-coupled stages, particular attention needs to be paid to bias. A negative feedback biasing system (see later) is usually required.

Capacitor coupling makes use of a capacitor placed in series between the output terminal of one stage and the input terminal of the next. The effect is that a.c. signals only can be coupled in this way, because d.c. levels cannot be transmitted through a capacitor. When amplifiers need a low −3 dB point when input frequency is only a few hertz, large values of capacitance will be required.

Transformer coupling makes use of current signals flowing in the primary winding of a transformer connected into the collector circuit of a transistor to induce voltage signals in the secondary winding, which in turn is connected to the base of the next transistor. Once again, only a.c. signals can be so coupled; and a well-designed transformer will be needed if signals of only a few hertz are to be coupled. Note that the gain/frequency graph of a transformer-coupled amplifier can show unexpected peaks or dips caused by resonances. For that reason, transformer-coupled amplifiers are seldom used when an even response is of importance.

Negative feedback

Negative feedback can be used in amplifier circuits either to stabilize bias (d.c. feedback) or to stabilize gain (a.c. feedback), or both. The conventional bias circuit illustrated earlier is a form of d.c. negative feedback, and where stages are directly coupled (as in operational amplifiers; see later), d.c. feedback over several stages will be needed to stabilize bias. Fault-finding bias problems can be very difficult in such circuits because every component in the feedback path has an effect on bias. For the moment, however, we shall concentrate on the use of a.c. feedback.

Although it is possible to design single amplifier stages with fairly exact values of voltage gain (it is, for example, quite possible to design a single-stage amplifier with a voltage gain of exactly 29 times, if that should happen to be wanted), it is less easy to design multistage amplifiers that will give the precise voltage gain required. The reason is that the input and output resistances of transistors vary considerably, depending on the varying h_{fe} values of individual transistors, and that in any form of signal coupling the output resistance of one transistor forms a voltage divider with the input resistance of the next, so attenuating the signal.

The point is simply illustrated in Figure 34.13, in which R_1 symbolizes the output resistance of the first transistor and R_2 the input resistance of the second.

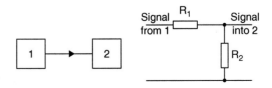

Figure 34.13 Attenuation of a signal when coupled from one stage to another

A more useful way of designing an amplifier for a specified figure of gain is to aim for one which has much too large a value of voltage gain, and then to use negative feedback to reduce this gain to the required figure. Negative feedback means taking a fraction of the output signal and applying it to the input of the amplifier in **antiphase**. One advantage of using negative feedback is that it is often possible to calculate the gain of the complete amplifier without knowing any of the individual transistor gains or resistances. Very often the gain of the amplifier is simply the ratio of two values of fixed resistors. Some feedback methods are shown in Figure 34.14, illustrating how the feedback signal can be added to the input signal.

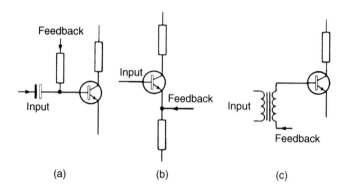

Figure 34.14   Some feedback methods, omitting bias and supply components. (a) Shunt feedback to base, (b) series feedback to emitter, (c) series feedback using a transformer

For negative feedback to be useful, the gain of the amplifier without feedback (called the **open-loop gain**) must be much greater (100 times or more) than the gain of the amplifier when feedback is applied (the **closed-loop gain**). When this situation exists, the **feedback factor**, (Closed-loop gain/Open-loop gain), becomes important, because distortion and noise generated in the transistors or other components within the amplifier will be reduced by exactly the same ratio. If, for example, closed-loop gain is 100 and open-loop gain is 10000, then the feedback factor becomes 100/10000, which is 1/100. This means that the distortion will be reduced to 1/100th of its value in the open-loop amplifier.

Bandwidth, in contrast, will be increased by the inverse of the factor, which is 100 times in this example, on the assumption that there is nothing else that will limit bandwidth in any other way. You cannot, for example, expect negative feedback to make the bandwidth of an amplifier greater than the bandwidth that the transistor(s) will handle.

Note that negative feedback cannot improve the performance of an amplifier for pulse waveforms if the **slew rate** of the amplifier is being exceeded. This topic will be dealt with later as part of the limitations of operational amplifiers.

Practical 34.3

Test and diagnose a fault in a feedback amplifier.

Operational amplifiers

Operational amplifiers (commonly known as op-amps) were originally developed to perform the equivalent of mathematical operations in analogue computers. Until integrated circuits were invented, however, their use in discrete form was very limited. Once mass production of integrated circuits started, op-amps were found to have many very useful properties for other applications. As a result, op-amps are now found in a very wide variety of equipment.

Integrated circuit op-amps are multistage differential amplifier units with direct coupling and very large values of gain. A differential amplifier is one that has two inputs, and it is the difference between the inputs that is amplified. The open-loop gain is the amount of gain that would be obtained by using the IC as an amplifier without feedback, but this amount is so large (usually 100 dB or more, corresponding to voltage gain of 100 000 or more) as to be unusable. Integrated circuit amplifiers are always used with negative feedback, and since direct coupling is normally used, the feedback is used both to establish bias and to establish signal gain. The advantages are:

- increased reliability due to a construction that leads to fewer interconnections to develop dry joints
- reduced size compared with their discrete-component counterparts
- reduced costs with volume production
- faster operation or better high-frequency response due to shorter signal path lengths
- IC fabrication gives a better control over the spread or variation of device parameters
- reduced assembly time due to fewer soldered joints.

Since the final performance of the circuit can be closely controlled by using negative feedback, the overall design of circuits is simpler.

A typical op-amp example, now quite old, is the 741. The remainder of this section will describe the operation of the IC version of this circuit.

The ideal op-amp would have the following characteristics:

- very high input resistance
- very low output resistance
- very high voltage gain, open-loop
- very wide bandwidth.

Table 34.3 shows typical characteristics for a 741 op-amp. You can see that the first three of the above requirements are easily met, while the bandwidth, although small by audio standards, is wide enough for the signals that the 741 was designed to handle, signals ranging from d.c. to a few hundred hertz.

Table 34.3 741 op-amp characteristics				
Input resistance	1M	Open-loop gain	100 dB	Max. supply voltage	±18 V
Output resistance	150R	Bandwidth	1 kHz @ 60 dB gain	Max. load current	10 mA

The gain–frequency response of a typical op-amp is shown in Figure 34.15. Remember that by using negative feedback the stage gain of an amplifier is reduced, but at the same time its bandwidth is increased. This is due to the fact that the amplifier's gain bandwidth product is a constant. This feature of being able to trade gain for bandwidth is probably the most important feature of the op-amp and forms one of the design steps in the production of a practical amplifier, but care is needed because there are limits imposed by slew rate, as we shall see.

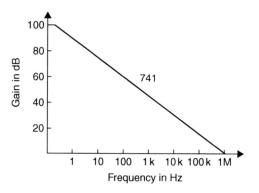

Figure 34.15   Typical op-amp gain–frequency graph

The frequency range of an op-amp depends on two factors, the **gain-bandwidth product** for small signals, and the **slew rate** for large signals. The gain-bandwidth product is the quantity $G \times B$, with G equal to voltage gain (*not* in dB) and B the bandwidth upper limit in hertz. For the 741, the GB factor is typically 1 MHz, so that, in theory, a bandwidth of 1 MHz can be obtained when the voltage gain is unity, a bandwidth of 100 kHz can be attained at a gain of 10, a bandwidth of 10 kHz at a gain of 100 times, and so on. This trade-off is usable only for small signals, and cannot necessarily be applied to all types of op-amp. Large amplitude signals are further limited by the slew rate of the circuits within the amplifier.

The **slew rate** of an amplifier is the value of change of output voltage per unit time: Slew rate = (Maximum voltage charge/Time taken), and the units are usually volts per microsecond. Because this rate cannot be exceeded for a given design of op-amp, the bandwidth of the op-amp for large signals, sometimes called the **power bandwidth**, is less than that for small signals.

The slew rate limitation cannot be corrected by the use of negative feedback; in fact, negative feedback acts to increase distortion when the slew rate limiting action starts, because the effect of the feedback is to increase the rate of change of voltage at the input of the amplifier whenever the rate is limited at the output. This accelerates the overloading of the amplifier, and can change what might be a temporary distortion into a longer lasting overload condition.

Slew rate limiting arises because of internal stray capacitances which must be charged and discharged by the current flowing in the transistors inside the IC, so that improvement is obtainable only by redesigning the internal circuitry. The 741 has a slew rate of about 0.5 V/μs, corresponding to a power bandwidth of about 6.6 kHz for 12 V peak sine-wave signals. The slew rate limitation makes op-amps unsuitable for applications that require fast-rising pulses, so a 741 should not be used as a signal source or feed (interface) with digital circuitry, particularly transistor–transistor logic (TTL) circuitry, unless a Schmitt trigger stage is also used. Higher slew rates are obtainable with more modern designs of op-amp; for example, the Fairchild LS201 achieves a slew rate of 10 V/μs.

The circuit symbol for an op-amp is illustrated in Figure 34.16(a). Two power supplies are needed (although single supply lines can also be used by modifying the circuits). The dual supply is balanced about earth. The op-amp has two inputs, labelled (+) and (−), and a single output. The internal circuit is that of a balanced amplifier, so the output voltage is an amplified copy of the voltage difference between the two inputs. The (+) sign at one input indicates that feedback from the output to this input will be in phase, positive. The (−) sign at the other input indicates that feedback to this input will be out of phase, negative.

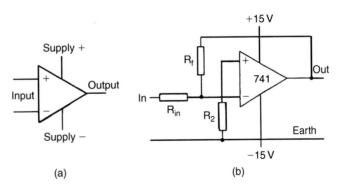

Figure 34.16 (a) Op-amp symbol, and (b) basic phase-inverting amplifier circuit

The voltage gain, open-loop, is some 10^5 (100 dB) or more. Since the internal circuit is completely d.c. coupled, both d.c. and a.c. negative feedback will be needed for linear amplifier applications.

An op-amp, unless the bias is stabilized, will suffer from **drift** and **offset**. Consider an op-amp with equal positive and negative power supplies. If both inputs are at precisely the same zero-voltage level, you would expect the output also to be at zero (ground) level. In fact, it will not be, and the difference between input levels needed to obtain a zero output is called the **offset**. It might be possible to apply inputs that ensured a steady output (even if not zero), but in such conditions the output voltage would vary with temperature and time, the effect called **drift**. The problems of offset and drift are dealt with by using d.c. feedback bias circuits. See later for more details of drift and offset in practical circuits.

The diagram of Figure 34.16(b) illustrates a basic phase-inverting amplifier circuit. Balanced power supplies are used, with the $(+)$ input earthed and the $(-)$ input connected to the output by a resistor R_f, which provides negative feedback of both d.c. and signal voltages. R_{in} serves to increase the input resistance. This inverting voltage amplifier circuit is the most common application for an op-amp.

Although the $(-)$ input is not actually connected to earth, its voltage (either d.c. or signal) is earth voltage, and it is said for this reason to be a **virtual earth**. What happens is this. The $(+)$ input is earthed, and any voltage difference between the inputs is amplified. Suppose, for example, that the $(-)$ input is at 0.1 mV below earth voltage. This 0.1 mV will be amplified by 10^5, giving a $+10$ V output (note the phase change). This voltage then drives current through R_f to increase the voltage at the $(-)$ input until it equals the voltage at the $(+)$ input.

Similarly, a 0.1 mV positive voltage at the $(-)$ input would cause the output to swing to -10 V, again causing the input to return to zero. Two points should be noted:

- The voltage gain is so high that the assumption that the $(-)$ input is at zero voltage is justified.

- Any unbalance in the internal circuit will cause an offset voltage at the input, so that the $(-)$ input may need to be at a very small positive or negative voltage to maintain the output at zero (the input offset voltage).

This input offset can be reduced in the type 741 by adding the offset balancing resistor shown in the circuit of Figure 34.17(a), in which the small

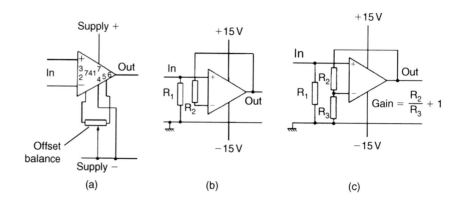

Figure 34.17   Op-amp circuits: (a) offset adjustment, (b) follower circuit, and (c) follower with gain

figures are the IC pin numbers. The size of the offset voltage will, however, vary as the temperature of the IC varies, so that some compensation may be needed in circuits such as high-gain d.c. amplifiers, in which offset can be troublesome. This is known as the input offset voltage temperature drift, usually abbreviated to drift. The offset is not a problem when large amounts of d.c. negative feedback are used.

Owing to the virtual earth at the $(-)$ input of the basic phase-inverting circuit, the circuit input resistance becomes equal to R_{in}. The figure for stage gain is simply the ratio R_f/R_{in}. To minimize the effects of temperature drift, the resistor R_2 is made equal to the parallel combination of R_f and R_{in}; that is,

$$\frac{R_f \times R_{in}}{R_f + R_{in}}$$

Again due to the virtual earth, several input signals can be applied to the $(-)$ input through resistors. This forms the basis of the summing amplifier, where the input signals are added.

Figure 34.17(b) shows a type of non-inverting circuit of the general type known as a **follower**. In this circuit the signal input is applied to the $(+)$ input, whose d.c. value is normally fixed at earth voltage by R_1. The feedback resistor R_2 ensures that the $(-)$ input is a virtual earth. A signal coming into the input will now produce an identical signal, in phase, at the output. The input resistance of the circuit is extremely high, the output resistance very low, and the circuit is used mainly to obtain these characteristics (like the emitter-follower transistor amplifier).

Unlike the familiar emitter-follower, this type of circuit can be modified to produce voltage gain, as shown in Figure 34.17(c). Such a circuit should only be used when the signal input is of small amplitude, because the effect of the feedback is to make the input signal a common-mode signal, and the amplitude of common-mode signals needs to be kept below a specified value in most op-amp designs. The gain in the example is given by:

$$G = \frac{R_2}{R_3} + 1$$

The characteristics of the op-amp make it possible to build integrating and differentiating circuits that give much better performance than can be obtained using only passive components.

Figure 34.18(a) shows a simple op-amp integrator which functions also as a low-pass filter. The time constant CR_2 seconds should be approximately five times the periodic time t of the input signal. A fully practical circuit, however, must also include some method of setting the output voltage to zero before the circuit is used, particularly if the d.c. level of the output is important. If the integrator is to be used for a.c. only, a bias resistor (such as R_1) will serve to prevent drift.

If C and R_2 of the integrator example are interchanged and their time constant is approximately one-fifth of the periodic time t, the circuit

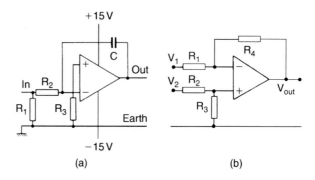

Figure 34.18 Op-amp circuits: (a) integrator, and (b) differential amplifier

becomes a differentiator. The output waveform then depends on the rate of change of the input signal. For input signals of a different shape, the differentiator acts as a high-pass filter.

When an op-amp is operated in the differential mode as drawn in Figure 34.18(b) (with bias omitted), its output, V_{out}, is proportional to the difference between the two inputs, V_1 and V_2. The polarity of V_{out} depends on which input is the larger. If V_1 is the greater, then V_{out} is negative.

Figure 34.19 shows positive feedback being used in an op-amp circuit. This has the effect of providing a Schmitt trigger action with hysteresis. If we imagine the input voltage starting low, then with input voltage rising, the output will suddenly switch over (high). As the input voltage falls, the output voltage will switch back again, but not at the same voltage as caused the switch in the other direction.

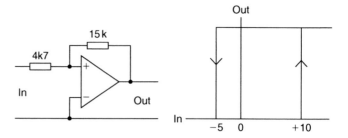

Figure 34.19 Op-amp providing Schmitt trigger action (power supplies omitted)

Op-amps can thus be used in many roles, the following being typical:

- as differential amplifiers, having exceptionally good common-mode rejection ratios, and with response down to d.c.
- as amplifiers for audio frequencies whose gain is controlled entirely by the values of the feedback components
- as high-pass and low-pass active filters

- as integrators and differentiators for the shaping of signal waveforms
- as sensitive switching circuits, triggered by the very small size of the differential input voltage needed to switch the output from one level to another
- as comparators, providing an output which is an amplified version of the voltage difference between two signal inputs (a.c. or d.c.). This application can be extended to provide a Schmitt trigger action with hysteresis.

Practical 34.4

Test and diagnose a fault in an operational amplifier circuit.

Power amplifiers

Both voltage amplifiers and current amplifiers play important parts in electronic circuits, but not all of them are capable of supplying some types of load. For instance, a voltage amplifier with a gain of 100 times may be well able to feed a 10 V signal into a 10 K load resistance, but quite incapable of feeding even a 1 V signal into a 10 ohm load. A current amplifier with a gain of 1000 may be excellent for amplifying a 1 μA signal into a 1 mA signal, but unable to convert a 1 mA signal at 10 V into a 1 A signal at the same voltage.

The missing factor common to both these examples is **power**. A signal of 1 A (r.m.s.) at 10 V (r.m.s.) represents a power output of 10 W, and the small transistors that are used for voltage or current amplification cannot handle such levels of power without overheating.

Transistors intended to pass large currents at voltage levels of more than a volt or so must have the following characteristics:

- low output resistance
- good ability to dissipate heat.

A low output resistance is necessary because a transistor with a high output resistance will dissipate too much power when large currents flow through it. Low resistance is achieved by making the area of the junctions much larger than is normal for a small-signal transistor. The ability to dissipate heat is necessary so that the electrical energy that is converted into heat in the transistor can be easily removed. If it were not, the temperature of the transistor (at its collector-base junction) would keep rising until the junctions were permanently destroyed.

Given suitable transistors with large junction areas and good heat conductivity to the metal case, the problem of power amplification becomes one of using suitable circuits and of dissipating the heat from the transistor.

In practice, most power amplifier stages are required to provide mainly current gain, since the voltage gain can be obtained from low-current stages before the power amplifier stage. The power amplifier stage can therefore have very low voltage gain or even attenuate the voltage level.

Heat sinks

Heat sinks take the form of finned metal clips, blocks or sheets which act as convectors passing heat from the body of a transistor into the air. Good contact between the body of the transistor and the heat sink is essential, and silicone grease (also called heat-sink grease) greatly assists in this contact.

Many types of power transistor have their metal cases connected to the collector terminal. It is therefore necessary to insulate them from their heat sinks. This is done by using thin mica washers between the transistor and its heat sink, with insulating bushes inserted on the fixing bolts. Mica is an electrical insulator and a heat insulator, but for a thin washer the heat conductivity can be reasonably good. Heat-sink grease should always be smeared on both sides of all such washers.

Classes of bias

Several different methods exist for biasing transistors that are to be used in power output stages. Class A and class B are the names used for two types of commonly used biasing systems for audio, and class C is used for RF transmitter amplifiers.

The transistor in a class A stage is biased so that the collector voltage is never bottomed, nor is current flow ever cut off. Output current flows for the whole of the input cycle. It is the bias system that is used for linear voltage amplifiers. Class A operation of a transistor ensures good linearity, but suffers from two disadvantages:

*   A large current flows through the transistor at all times, so that the transistor needs to dissipate a considerable amount of power.
*   This loss of power in the transistor inevitably means that less power is available for dissipation in the load, and a class A stage can seldom be more than about 30% efficient, meaning that the a.c. power can seldom be more than 30% of the d.c. power.

Even in an ideal class A amplifier, with the load and amplifier output resistances matched, only some 50% of the available power would be delivered to the load, with the remaining 50% being dissipated by the amplifier in the form of heat. When the resistances are mismatched, the power transfer ratio is even lower.

In a class B amplifier (Figure 34.20), the power transistor conducts for only one-half of the duration of the input sine wave. A single transistor is therefore unusable unless the other half of the wave can be obtained in some other way. At radio frequencies this can be done by making use of a load which is a resonant circuit (called a **tank circuit**). The tank circuit is made to oscillate by the conduction of the transistor, and the action of the resonant circuit continues the oscillation during the period when the transistor is cut off.

This principle can only be applied, however, if the signal to be amplified is of fairly high frequency. This is because the values of inductance and capacitance needed to produce resonance at, say, audio frequencies would be too large to be practical, and because a load of such a size would in any case restrict the bandwidth too much.

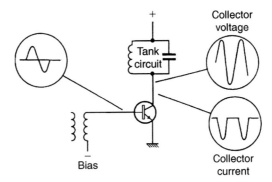

Figure 34.20 Class B radio-frequency amplifier

An alternative method, which is used at audio and other low frequencies, is to use two transistors, each conducting on different halves of the input wave. Such an arrangement is called a **push–pull circuit** (Figure 34.21). Push–pull circuits can be used in class A amplification, and are essential for use in class B audio amplifiers.

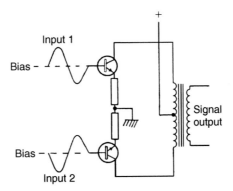

Figure 34.21 Simple push–pull circuit using a transformer to combine the antiphase outputs from the transistors

Class B stages have the following advantages over class A:

- Very little steady bias current flows in them, so that the amplifier has only a negligible amount of power to dissipate when no signal is applied.
- Their theoretical maximum efficiency is 78.4%, and practical amplifiers can achieve efficiencies of between 50 and 60%. This means that more power is dissipated in the load itself, and less wasted as heat by the transistors.

The disadvantages of Class B compared to Class A are that:

- The supply current changes as the signal amplitude changes, so that a regulated supply is often needed.

- More signal distortion is caused, especially at that part of the signal where one transistor cuts off and the other starts to conduct. This part of the signal is called the cross-over region.

The circuit of Figure 34.22(a) illustrates a single-transistor power amplifier biased in class A. Because no d.c. can be allowed to flow in the load (which is in this case a loudspeaker), a transformer is used to couple the signal from the transistor to the load.

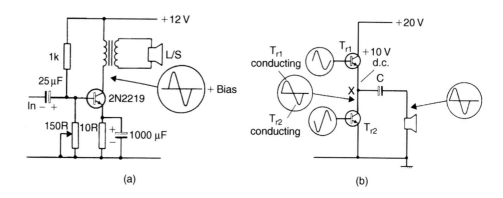

Figure 34.22   (a) Simple class A stage, and (b) the single-ended push–pull circuit

With no signal input, the steady current through the transistor is about 50 mA and the supply voltage is 12 V. Because of the low resistance of the primary winding of the transformer, the voltage at the collector of the transistor is about equal to the supply voltage.

When a signal is applied, the collector voltage swings below supply voltage on one peak of the signal and the same distance above the supply voltage on the other peak of the signal. Because of this transformer action the average voltage at the collector of the transistor remains at supply voltage when a signal is present.

In a class A stage of this type the current taken from the supply is constant whether a signal is present or not. Any variation in current flow from no-signal to full-signal conditions indicates that some non-linearity, and therefore distortion, must be present.

A circuit type that is often used for class B amplifiers is illustrated in Figure 34.22(b). It is known as the **single-ended push–pull** or **totem-pole** circuit. Two power transistors are connected in series, with their mid-connection (point X in the figure) coupled through a capacitor to the load, which may typically be a loudspeaker, the field coils of a television receiver or the armature of a servomotor. During the positive half of the signal cycle Tr_1 conducts, so that the output signal drives current through the load from X to ground. During the negative half of the cycle, Tr_1 is cut off and Tr_2 conducts, so that the current now flows in the opposite direction, through the load and Tr_2.

The coupling capacitor C forms an essential part of the circuit. When there is no signal input, C is charged to about half supply voltage, i.e. the voltage at point X. The voltage swing at this point X, from full supply voltage in one direction to ground or zero voltage in the other, thus becomes at the load a voltage swing of the same amplitude centred around zero volts.

Bias and feedback

Class A output stages need a bias system that will keep the standing quiescent current (i.e. the d.c. current with no signal input) flowing through them almost constant, despite the large temperature changes caused by the standing current. One common method is to use a silicon diode as part of the bias network. The diode should be a junction diode attached to the same heat sink as the power transistor(s).

As a silicon junction is heated, the junction voltage (about 0.6 V at low temperatures) that is needed for correct bias becomes less. A fixed-voltage bias supply composed of resistors would therefore overbias the transistor as the temperature increased. A silicon diode compensates for the change in the base-emitter voltage of the transistor, since the forward voltage of the diode is also reduced as its temperature rises.

The class A push–pull stage in Figure 34.23 uses two silicon transistors coupled to the load by a transformer. The input signal is also coupled to the stage by a transformer, which ensures that both transistors obtain the correct phase of signal. The use of a second transformer in such a circuit is often undesirable because it reduces the bandwidth and makes the amplifier less stable if negative feedback is used (because of the unpredictable changes

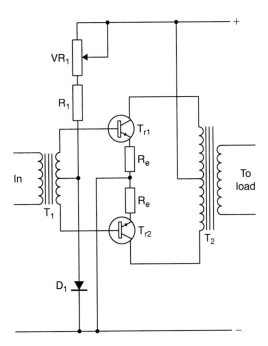

Figure 34.23 Class A push–pull circuit

of phase that occur in a transformer at extremes of frequency). A phase-splitter stage that uses transistors rather than a transformer can be used. Such a phase-splitter can use a transistor with equal loads in collector and emitter, or a long-tailed pair circuit with unbalanced input and balanced output.

The bias current for this circuit is taken to the centre-tap of the secondary winding of the phase-splitter (or driver) transformer T_1, and some additional stability is obtained by using the negative feedback resistors R_e in the emitter leads. Any increase in the bias current of the power transistors will cause the emitter voltage of both transistors to rise, so reducing the voltage between base and emitter and thereby giving back-bias to the input.

In both the circuits that have been described so far the bias is adjusted so that the correct amount of steady bias current flows in the output stage. To adjust this bias current in the single-transistor stage requires breaking the collector circuit to connect in a current meter, and then adjusting a potentiometer so as to give the correct current reading. The push–pull circuit can be set by measuring the voltage across the emitter resistors, R_e, and by adjusting VR_1 to give the correct value of voltage at these points.

A more elaborate circuit using the **class AB** complementary stage is shown in Figure 34.24. In class AB, each amplifier is biased to a value lying between those appropriate for either class A or class B. The output current in each stage thus flows for slightly more than half of each input cycle. The effect is to minimize the cross-over distortion that occurs in class B operation when both transistors are non-conducting.

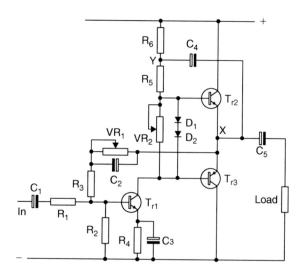

Figure 34.24 Complementary class AB output stage of the totem-pole type with driving stage and feedback connections

In this circuit, both a.c. and d.c. feedback loops are used. The d.c. feedback is used to keep the steady bias current at its correct low value, and the a.c. feedback is used to correct the distortions caused by class AB operation, in particular the cross-over distortion.

Two bias-adjusting settings are needed in this circuit. The potentiometer VR_1 sets the value of the bias current in Tr_1, so that the amount of current flowing through resistors R_2, R_3 and VR_2 is controlled. The potentiometer is adjusted so that the voltage at point X is exactly half the supply voltage when there is no signal input. Potentiometer VR_2 controls the amount of current passing through the output transistors.

This type of output circuit is called a **complementary stage**, because it uses complementary transistors, one NPN and one PNP type, both connected as emitter followers. With the emitters of both Tr_2 and Tr_3 connected to the voltage at point X, both output transistors are almost cut off when no signal is present. When VR_2 is correctly adjusted, the voltage drop between points A and B is just enough to give the output transistors a small standing bias current (2–20 mA) to ensure that they never cut off together. The value of this steady current is usually set low, to keep cross-over distortion to a minimum. The actual value will be that recommended by the manufacturers. A current meter must be used to check the value of bias current. Another name for this circuit is the **Lin** circuit, named after its inventor.

You can see that there are two a.c. feedback loops in this amplifier. Negative feedback, to improve linearity, is taken through C_2, and positive feedback is taken through C_4 to point Y. This positive feedback, sometimes called **bootstrapping**, cannot cause oscillation in this circuit because the feedback signal is always of smaller amplitude than is the normal input signal at that point (an emitter follower has a voltage gain slightly lower than unity). It has, however, the desirable effect of decreasing the amount of signal amplitude that is needed at the input to Tr_2.

Impedance matching

The ideal method of delivering power to a load would be to use transistors which had a very low resistance, so that most of the power (I^2R) was dissipated in the load. Most audio amplifiers today make use of such transistors to drive 8 ohm loudspeaker loads.

For some purposes, however, transistors that have higher resistance must be used or loads that have very low resistance must be driven, and a transformer must be used to match the differing impedances. In public address systems, for example, where loudspeakers are placed at considerable distances from the amplifier, it is normal to use high-voltage signals (100 V) at low currents so as to avoid I^2R losses in the lines. In such cases, the 8 ohm loudspeakers must be coupled to the lines through transformers.

To provide the maximum transfer of power, the turns ratio, N, of the transformer must be:

$$N = \sqrt{\frac{\text{Output impedance of amplifier}}{\text{Impedance of load}}}$$

Faults in output stages

Faults in output stages are usually caused by overdissipation of power, which in turn can be caused by overloading or overheating. An output stage can be overloaded, when capacitor coupling is used, by connecting a load having too low a resistance. The usual result is to burn out the output transistor(s).

In most class AB totem-pole circuits, even the most momentary short-circuit at the output (caused, for example, by faulty connections) will cause the transistor that is connected to the positive supply line (Tr_2 in Figure 34.24) to burn out if a signal is being amplified.

Excessive bias currents can often be traced to the failure of a diode in the bias chain, or to a burnout of the biasing potentiometer in a totem-pole circuit.

Unexpected clipping of the output can be caused by failure of the boot-strap capacitor (C_4 in Figure 34.24), or by a fault in the bias resistors which has caused the voltage at point X to drift up or down.

Integrated circuit power amplifiers are now widely used, replacing power stages that use separate (discrete) transistors. Unlike op-amps, however, power amplifier ICs take no standardized form, although many use the same scheme of connections. Figure 34.25 shows a circuit diagram (courtesy of RS Components Ltd) for the TDA2030 power amplifier used with a single 15 V power supply. The IC is manufactured on a steel tab which can be bolted to a heat sink for efficient cooling. Although this is now an old variety of IC, the principles are similar, and this age of component is likely to appear in equipment brought in for servicing.

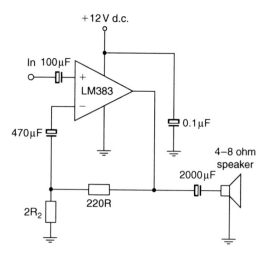

Figure 34.25 An IC power amplifier circuit (courtesy of National Semiconductor Inc.) using the LM383 IC

The circuit is very similar to that of a non-inverting op-amp voltage amplifier, but the practical layout of the circuit has to be designed so that the decoupling capacitors (C_2 and C_3) are very close to the IC. The electrolytic capacitor C_3 needs to be bypassed by a plastic dielectric capacitor because the impedance of an electrolytic capacitor rises at the higher frequencies. At these frequencies, C_2 performs the decoupling in place of C_3. The series circuit of R_6 and C_6 is designed to maintain stability and is called a **Zobel network**; it is also widely used in discrete transistor amplifiers. This circuit will deliver up to 13 W of audio output into the 4 ohm loudspeaker.

Note that the circuits used in hi-fi amplifiers are usually of a specialized nature, and many use power MOSFETs in the output stages. Some types can be serviced, using information from the manufacturers, but others should be returned to the manufacturer for servicing, particularly if there has been a failure of any critical part. For example, failure of one power MOSFET will require replacement with a matching unit, and few servicing workshops will have equipment for precisely measuring power MOSFET parameters.

Multiple-choice revision questions

34.1 Increasing the value of the load resistance on a single-stage voltage amplifier will:
   (a) increase the bias current
   (b) decrease the bias current
   (c) increase the gain
   (d) increase the bandwidth.

34.2 Feedback of signal from the collector of a transistor directly to its base would cause:
   (a) reduced gain
   (b) more distortion
   (c) oscillation
   (d) reduced bandwidth.

34.3 A transformerless push–pull output stage might use:
   (a) a stabilized power supply
   (b) a phase splitter
   (c) a Lin circuit
   (d) high-resistance transistors.

34.4 A transistor amplifier is to be used at d.c. and very low frequencies. It *cannot* therefore use:
   (a) negative feedback
   (b) capacitor coupling
   (c) a differential circuit
   (d) balanced power supplies.

34.5 An ideal op-amp would have (among other things):
   (a) very high output impedance
   (b) very high input impedance
   (c) very low gain
   (d) very low common-mode rejection ratio.

34.6 At a virtual earth point you would expect to find:
   (a) negligible d.c. voltage
   (b) negligible feedback
   (c) negligible offset voltage
   (d) negligible a.c. voltage.

35   Oscillators and waveform generators

Oscillators are critical to many systems: they act as clock sources for digital systems, carrier signal sources for radio transmitters, and sweep generators for television and oscilloscopes. Oscillators are usually built from amplifier stages and filter (delay) stages, using feedback to sustain the signal.

Positive feedback

Negative feedback is achieved by subtracting a fraction of the output signal from the input signal of an amplifier. In practice, this is done by adding back the feedback signal in **antiphase**, so that feedback from an output which is in antiphase to an input is always negative unless some change of phase occurs in the circuit used to connect the output to the input (see later, under Oscillators for low frequencies).

If a signal which is *in phase* with the input is fed back, the feedback is positive. Positive feedback takes place when a fraction of the output signal is added to the signal at the input of an amplifier, so increasing the amplitude of the input signal. The result of positive feedback is higher gain (although at the cost of more noise and distortion) if the amount of feedback is small. If the amount of feedback is large, the result is oscillation.

An amplifier will oscillate when both of the conditions below are met:

- The feedback is positive at some frequency.
- The voltage gain of the amplifier is greater than the attenuation of the feedback.

For example, if 1/50th of the output signal of an amplifier is fed back in phase, oscillation will take place if the gain (without feedback) of the amplifier is more than 50 times, and if the feedback is still in phase.

Oscillator feedback circuits are arranged so that only one frequency (called the fundamental) of oscillation is obtained. This can be done by ensuring that:

- the feedback is in phase at only one frequency
- the amplifier gain exceeds feedback loop attenuation at one frequency only, or
- the amplifier switches off entirely between timed conducting periods.

Oscillator circuits are of two types. Sine-wave oscillators use the first two methods above for ensuring constant frequency operation. **Aperiodic** (or untuned) oscillators, such as multivibrators, make use of the third method. Oscillators are equivalent to amplifiers which provide their own inputs. They also convert the d.c. energy from the power supply into a.c.

Some oscillator circuits operate in **class C** conditions. Even if the transistors start off with some bias current flowing, the action of the oscillator will turn off the bias for quite a large proportion of the complete waveform, making the transistor operate in class C once it is oscillating. The reason for this will be clearer as you read through this chapter; it arises because a tuned circuit connected to an oscillator will continue to oscillate for a short time even when the transistor is no longer conducting. Aperiodic oscillators all operate in class C.

Sine-wave oscillators

A sine-wave oscillator consists of an amplifier, a positive feedback loop and a tuned circuit which ensures that oscillation occurs at a single definite frequency. In addition, there must be some method of stabilizing the amplitude of the oscillations so that the oscillation neither stops nor builds up to such an amplitude that the wave becomes distorted by reason of bottoming or cut-off.

The most common types of sine wave oscillator are those which operate at radio frequencies, such as are used in the local oscillators for superhet receivers, which use LC tuned circuits to determine the oscillating frequency.

Like most oscillator circuits, the **Hartley oscillator** exists in several forms, but the circuit illustrated in Figure 35.1(a) is a much-used type. The tuned circuit L1C2 has its coil tapped to feed a fraction of the output signal back through C4 to the emitter of Q1. Since an output at the emitter is always in phase with the output at the collector, this feedback signal is positive. The base voltage of Tr1 is fixed by the values of the resistors R1 and R2, with C1 acting as an a.c. bypass capacitor.

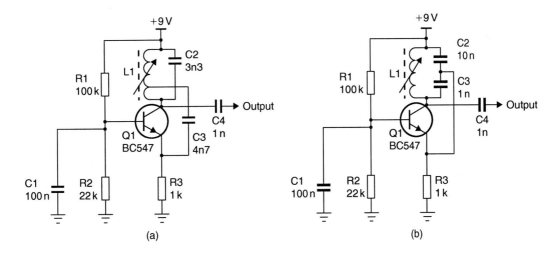

Figure 35.1 Radio frequency LC oscillators: (a) Hartley, and (b) Colpitts

The amplitude of the oscillation is limited at the emitter because the transistor will cut off if the emitter voltage rises to a value more than about 0.5 V below base voltage. The distortion of the waveshape caused by this limiting effect is smoothed out by the flywheel effect of the tuned (or tank) circuit L1C2, which produces a sine-wave voltage at the resonant frequency even when the current waveform is not a perfect sine wave.

Irrespective of how its feedback is arranged, the Hartley oscillator can always be recognized by its use of a tapped coil. Its frequency of oscillation, as is the case with all oscillators using LC tuned circuits, is given by the formula:

$$f = \frac{1}{2\pi\sqrt{LC}}$$

Faults that can cause failure in an oscillator of this type include the following:

- bias failure caused by breakdown of R1, R2 or R3
- a faulty bypass capacitor
- a leaky or open-circuit coupling capacitor C4
- faults in either C2 or L1.

C2 should be a high Q, low loss part, typically polystyrene or multilayer ceramic using a high Q dielectric (C0G or NP0). In some older equipment silver-mica type capacitors were used, but these are now mostly obsolete and are not available in surface-mounted versions. Some low Q capacitors (X7R or Y5V dielectric) will not permit oscillation because of losses, meaning that the capacitor dissipates so much power that the loop gain falls below unity. More often, lossy capacitors degrade the frequency stability of the oscillator.

The **Colpitts oscillator** (Figure 35.1b) is basically similar to the Hartley oscillator. Instead of using a tapped coil, however, the Colpitts oscillator uses the combination of C2 and C3 to tap off a fraction of the output voltage to feed back into the base. The latter is biased and bypassed in the same way as in the Hartley circuit.

The same remarks about circuit operation, and about the several possible circuit configurations, apply to the Colpitts as to the Hartley oscillator, but the formula for determining the frequency of oscillation is slightly different. Because the capacitors C2 and C3 are in series, it is the series combination C' (in which $1/C' = 1/C2 + 1/C3$) that tunes L1 to give the output frequency. The formula must therefore use the value of C' rather than C2 or C3.

The oscilloscope is the most certain way of checking that an oscillator circuit is working, and also of measuring the frequency. Another check is to measure the current drawn by an oscillating circuit, which will be much larger than normal if the circuit is not oscillating (because the oscillator operates in class C when oscillating, requiring less bias current).

Practical 35.1

Construct the Colpitts oscillator of Figure 35.1(b). The inductor should be approximately 50 turns of 28 swg wire on a 10 mm ferrite core, or a 50–200 μH fixed inductor may be used. Use an oscilloscope to measure the signal at the collector of Q1. Measure the frequency of oscillation, adjust the power supply voltage to 12 V and measure the frequency of oscillation again. Warm the circuit gently with a hairdryer: what happens to the frequency of oscillation?

Tuned-load oscillators (Figure 35.2) have a tuned circuit as the load of the transistor and use another component for feedback. These oscillators are essentially variants of the Hartley oscillator described previously.

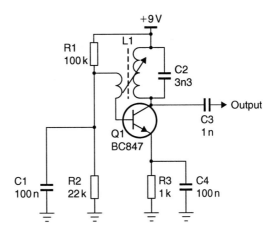

Figure 35.2   Tuned load oscillator

Remember that the output of a single-transistor common-emitter amplifier is always phase inverted, so that another inversion must be carried out if feedback is to be taken from the collector to the base. One way of achieving this is to use a coupling winding on the same former or close to the collector inductor, connected to give a phase inversion. Amplitude limiting is carried out in this circuit by the bottoming and cut-off action, and the output is smoothed into a sine wave by the resonant oscillations of the tuned circuit.

Failure of this type of oscillator circuit can be caused by the bypass capacitor C4 going open-circuit, as well as by any of the biasing faults that can cause the Hartley oscillator to fail.

Crystal oscillators make use of a quartz crystal in place of a tuned circuit, to give much greater frequency stability than can be achieved under the same conditions by any LC circuit. Crystal resonators can easily achieve loaded Q values in excess of 10 000, compared with a good LC resonator with a loaded Q of 100.

There is a great variety of crystal oscillator circuits, some of which use the crystal as if it were a series LC circuit, with others using it as part of a parallel LC circuit in which the crystal replaces the inductor. Figure 35.3 illustrates a form of Colpitts oscillator, but with the crystal providing the frequency-determining feedback path and some of the 180° phase shift between collector and base. The choke L1 acts as the collector load across which an output signal is developed and also contributes to the phase shift.

In normal use, crystal oscillators are extremely reliable, but excessive signal current flowing through the crystal can cause it to break down and fail. The usual comments concerning bias and decoupling components also apply here. Oscillators of the same type and performance can also make use of surface acoustic wave (SAW) filters to determine frequency.

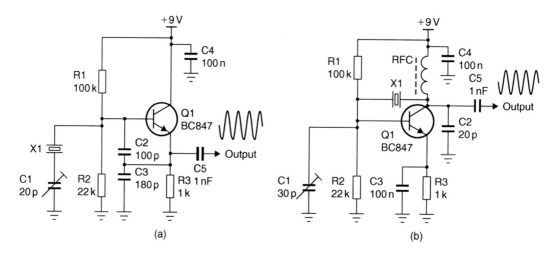

Figure 35.3   Crystal oscillators: (a) Colpitts, and (b) Pierce

Note that all the circuits for sine-wave oscillators can be constructed with metal oxide semiconductor field-effect transistors (MOSFETs) in place of bipolar transistors. The use of MOSFETs has several advantages, particularly where crystal oscillators are concerned, because no input current is required at the gate of a MOSFET. However, MOSFETs are inherently noisier than bipolar transistors, so in high-performance oscillators their use may not be desirable; this is one of the few remaining applications where junction field-effect transistors (JFETs) can have an advantage.

Oscillators for low frequencies

Oscillators that are required to operate at low frequencies cannot use LC tuned circuits because of the impractically large size of inductor that would be needed. An alternative construction is the RC oscillator, and Figure 35.4(a) shows the basic outline of one type, known as the **phase-shift oscillator** or **divider-chain oscillator**.

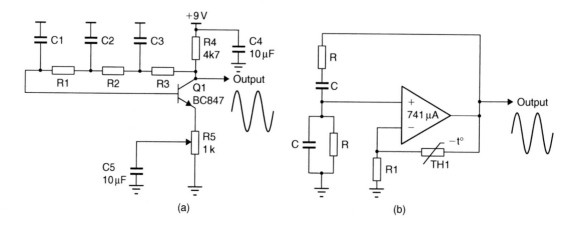

Figure 35.4   (a) Phase-shift oscillator, and (b) thermistor-stabilized Wien bridge circuit

Every RC potential divider must attenuate and shift the phase of the signal at the collector of the transistor. If the total phase shift at a given frequency is 180°, the signal fed back to the base will be in phase with the signal at the collector and the circuit will oscillate.

The output waveform will not, however, be a sine wave unless the gain of the amplifier can be so controlled that it is only just enough to sustain the oscillation. All RC oscillators therefore require an amplitude-stabilizing circuit, which is usually provided by a negative feedback network. This network usually includes a component such as a thermistor whose resistance decreases as the voltage across it increases.

In this way, an increase in signal amplitude causes an increase in the amount of negative feedback, which in turn causes the amplifier gain to decrease, so correcting the amplitude of oscillation. Failure in this feedback network will either stop oscillation or cause severe distortion of the output waveform.

Another type of RC oscillator is shown in Figure 35.4(b). This type of oscillator is known as the **Wien bridge**. Note that the circuit uses feedback to the non-inverting (in-phase) input of the amplifier.

All RC oscillator circuits require amplitude stabilization if they are to produce sine waves of good quality, but they are capable of providing very low distortion figures (in the order of only 0.01%) by good design. The outline example of a Wien bridge oscillator circuit with provision for amplitude stabilization by a thermistor is illustrated in Figure 35.5(b). The amplifier is a 741 integrated circuit (IC). A practical circuit would need to contain some setting to allow for adjustment of the feedback so that the thermistor could hold the amount of feedback at a level that would only just permit oscillation.

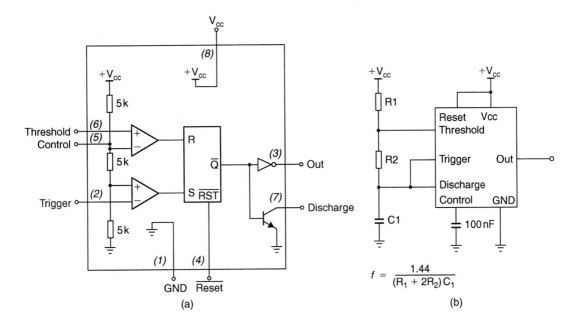

$$f = \frac{1.44}{(R_1 + 2R_2)\,C_1}$$

(a)   (b)

Figure 35.5 (a) Simplified internal schematic for the 555 timer, and (b) astable 555 circuit

A distorted output from an RC oscillator, or zero output, is nearly always caused by failure of a component in the amplitude-stabilizing circuit. Lack of output can also be due to a sudden loss of gain, such as could be caused by bias failure or by the failure of a decoupling capacitor.

Practical 35.2

Construct the phase shift oscillator shown in Figure 35.4(a). Use 220 k resistors for R1, R2 and R3, and 1 nF capacitors for C1, C2 and C3. Adjust R5 so that the circuit just oscillates. Use an oscilloscope to monitor the output waveform at the collector of Q1. Sketch the waveform as R5 is adjusted.

The 555 timer integrated circuit

The 555 timer is an IC that is widely used as an oscillator or as a generator of time delays. The timing depends on the addition of external components (a resistor and a capacitor), and one very considerable advantage of using the IC is that the time delays or waveforms that it produces are well stabilized against changes in the d.c. supply voltage. As with other ICs, we shall ignore the internal circuitry (other than a block diagram) and concentrate on what the chip does (Figure 35.5).

The chip pin connections for the usual eight-pin dual in line (DIL) form of the chip are shown in Figure 35.5(a), using pins 1 and 8 for earth and supply positive, respectively. The output is from pin 3, and the other pins are used to determine the action of the chip. Of these, pins 2, 6 and 7 are particularly important. Pin 7 provides a discharge current for a capacitor that is used for timing, and pin 6 is a switch input that will switch over the output of the circuit as its voltage level changes. For most uses of the chip, these pins are connected to a CR circuit, whose charge and discharge determine the time delay or the wavetime of the output.

An astable pulse generator circuit is illustrated in Figure 35.5(b). Pin 6 is used to discharge the capacitor C1 and is connected to the triggering pin 2. This ensures that the circuit will trigger over to discharge the capacitor when the voltage level reaches two-thirds of supply voltage, and will trigger back to allow the capacitor to charge when the voltage level reaches one-third of supply voltage. The frequency of the output can be adjusted by using a 100 K variable in place of R2 and, if needed, by using switched values of capacitance C1.

For the astable circuit illustrated, the waveform timings are given approximately by the following formulae:

Charge time:	0.7 (R1 + R2)C1
Discharge time:	0.7 R2C1
Periodic time:	0.7 (R1 + 2R2)C1
Frequency:	1.44/(R1 + 2R2)C1

Practical 35.3

Construct an astable oscillator using a 555 timer IC. Using the circuit and formulae for frequency given in Figure 35.5(b), select components that will give frequencies of 1 Hz, 1 kHz and 100 kHz, measure the output frequency of your circuit and comment on the component selection. Measure the supply current for each frequency.

The **voltage-controlled oscillator** (**VCO**) is usually a standard LC oscillator circuit whose output frequency can be controlled by an input voltage applied at one terminal. One way of implementing this is to use an astable circuit in which the rate of charge and discharge is controlled by a transistor rather than a resistor. For higher frequencies, the VCO is usually implemented by using a **varactor diode** (Figure 35.6) as part of the tuned circuit of an oscillator. Typically, a Colpitts circuit is used. The voltage control function allows the oscillator's output frequency to be controlled, either to make a frequency modulation (FM) modulator or by using a feedback control system to lock its frequency to a multiple of another frequency, using a phase-locked loop (PLL).

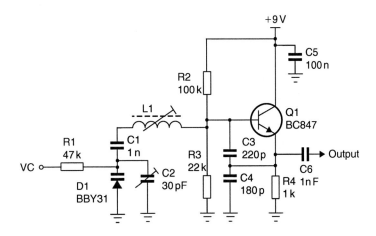

Figure 35.6   Voltage-controlled oscillator

The **PLL** is an important circuit which forms the basis of many important functions in consumer, industrial and other electronic systems. PLLs are used in television receivers to synchronize the local colour decoder, as well as in synthesized tuners to select the channel frequency.

Figure 35.7 shows the block diagram of the 4046, a common complementary metal-oxide semiconductor (CMOS) PLL IC. The heart of the circuit is an RC VCO circuit. The oscillator output is compared with that of an input signal, by one of several phase comparator circuits. If the difference between the VCO frequency and input signal frequency is within a certain range, the capture range of the loop, the loop locks and the oscillator frequency is pulled towards that of the input signal by the d.c. voltage generated by the phase detector.

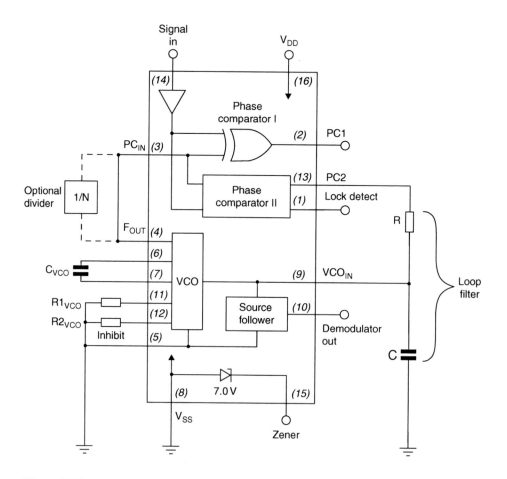

Figure 35.7   Block diagram of the 4046 PLL IC

The 4046 can be used simply as a precision VCO whose frequency is set by the values of the resistors R1 and R2 and capacitor C1, as well as the voltage on pin 9 (Figure 35.7). The output frequency is available at pin 4, the rest of the IC being disabled by connecting pins 3 and 14 to VSS and leaving pins 1, 2 and 13 unconnected. Alternatively, the PLL part of the 4046 can be used with an external VCO, the internal VCO being disabled by pulling the VCO inhibit pin, pin 5, high.

One common use of the 4046 is as an FM or frequency shift keying (FSK) detector. The internal oscillator frequency and tuning slope can be set over a wide range, and the bandwidth over which the internal oscillator will lock to the frequency of the incoming signal can be set by the choice of phase comparator and loop filter components. In this way the VCO can be made to track an incoming FSK or FM signal, with the original modulating signal being reproduced at the demodulate output pin, pin 10.

Using a divider circuit between pins 4 and 3 of the 4046 allows frequency multiplication by locking the VCO output divided by N to the input signal; for example, if the input to the PLL is a crystal-controlled square wave, the timing components can be set so that the output frequency is at N times the frequency of the crystal (provided that this does not exceed the frequency range of the VCO). The output frequency selected in this way will be perfectly locked in phase to the crystal frequency.

Such an arrangement is used in frequency-synthesizer circuits, since it can combine the versatility of the variable-frequency LC oscillator with the stability of crystal control. Component values for the loop filter, which determines how quickly the loop will lock and how the loop will follow a changing frequency input signal, can be calculated from the manufacturer's data sheet. These calculations are dependent on the phase detector being used and the exact details of the application.

A major advantage of PLL circuits is that they make it possible to regenerate a signal that is practically lost in noise. Provided the input signal amplitude is enough to drive the PLL (only about 180 mV is needed for the 4046 when operated from a 5 V supply), any input frequency that is within the capture range will be locked in, and the output will be a waveform of that frequency which is free from both noise and interference.

Sawtooth generators

The action of an integrating circuit to convert a square wave into a wave with sloping sides makes it the basis for oscillator circuits that generate a sawtooth waveform used for timebases. A simple timebase circuit is shown in Figure 35.8. While Q1 is conducting, the voltage at its collector is very

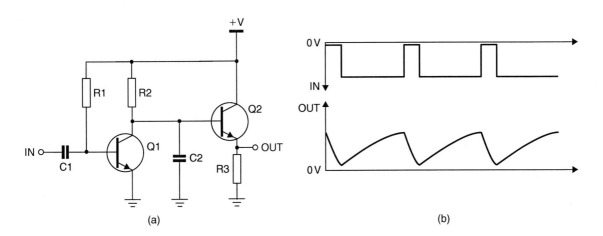

Figure 35.8   (a) Simple timebase or sweep circuit, and (b) waveforms

low and C2 is discharged. A negative-going pulse arriving (from an asymmetric astable multivibrator, for example) at the base of Q1 will cut off Q2, and C2 starts to charge through R2. This integrating action generates a slow-rising waveform which forms the sweep part of the sawtooth.

As the trailing edge of the pulse reaches its base, Q1 is switched on again. C2 rapidly discharges through Q1 and causes the rapid flyback at the end of the sweep waveform. Q2 is an emitter-follower acting as a buffer stage to prevent the waveform across C2 from being affected by the input resistance of circuits to which the sawtooth is coupled.

In such a simple circuit, the waveform across the capacitor will, in practice, be approximately a straight line if C2 is allowed to charge only to a small fraction of the supply voltage. Thereafter, the waveform will tend to bend over towards the horizontal. To prevent this, it is important to make the time constant R2C2 much longer than the duration of the square pulse applied to the base of Q1.

Timebase circuits normally use considerably more elaborate methods than that shown in Figure 35.8, with the object of ensuring that the sweep voltage remains linear. One type of sawtooth generator uses constant-current circuits to replace R2. Another, the Miller timebase, uses negative feedback to keep the sweep waveform more truly linear.

The **Miller integrator** is a circuit based on the application of negative feedback through a capacitor. In the circuit of Figure 35.9, C1 (the **Miller capacitor**) is connected between the collector and the base of a transistor. A rise of positive voltage at the input large enough to start the transistor conducting would, if C1 were not there, cause a large negative change in voltage at the collector. Negative feedback through C1, however, prevents the voltage at the base of Q1 from rising more rapidly than the rate at which C1 can charge through R1.

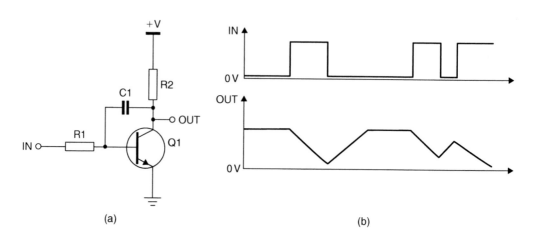

Figure 35.9   (a) Basic Miller integrator circuit, and (b) idealized waveforms

This charging rate does not follow the usual exponential shape (while the transistor is conducting) because of the feedback through C1, and a good linear sweep waveform is obtained. The principal application of the Miller integrator is in timebase circuits for oscilloscopes. If the gain of the transistor is high, the sweep waveform can be made very linear indeed.

A complete Miller timebase circuit for an oscilloscope is made considerably more elaborate than the basic circuit shown by the addition of other circuits whose purposes are (a) to discharge the Miller capacitor more rapidly (so obtaining rapid sweep flyback), and (b) to switch the input voltage at the times required to generate the timebase waveform.

Multiple-choice revision questions

35.1 What type of sine-wave oscillator uses a tapped inductor to provide the positive feedback required for oscillation?
(a) Clap
(b) Colpitts
(c) Hartley
(d) Pierce.

35.2 What type of capacitor is usually used in the tuned circuits of oscillators?
(a) electrolytic
(b) low loss
(c) high voltage
(d) surface mount.

35.3 An LC oscillator uses a $100\,\mu\text{H}$ inductor. What value of capacitor is required to tune it to $1.8\,\text{MHz}$?
(a) $16\,\text{pF}$
(b) $22\,\text{pF}$
(c) $47\,\text{pF}$
(d) $78\,\text{pF}$.

35.4 An RC phase-shift oscillator uses three identical RC stages in the feedback circuit. What phase shift is required in each stage?
(a) $30°$
(b) $60°$
(c) $90°$
(d) $120°$.

35.5 A timebase circuit uses a Miller integrator to produce the sweep signal for an oscilloscope. If the timebase circuit uses a bipolar junction transistor (BJT), where is the integrator capacitor connected?
(a) collector-emitter
(b) base-emitter
(c) collector-base
(d) base-positive supply.

Unit 4

Outcomes

1. Demonstrate an understanding of logic families and their terms used in their specifications
2. Demonstrate an understanding of time division multiplex (TDM)
3. Demonstrate an understanding of sequential logic
4. Demonstrate an understanding of common fault types and an understanding of faultfinding procedures and test equipment.

36 Logic families and terminology

Digital logic devices may be members of families of standardized types, or they may be tailor-made for a particular application. At one time, digital circuitry was composed almost entirely of standard functions, that is gates, latches, dividers, etc., making replacement easy. Manufacturing cost and volume requirements tended to drive towards the use of the minimum parts count and hence application-specific integrated circuits (ASICs) that are made specifically to carry out all, or at least a large part of the required circuit functions. These ASICs are often hard to obtain as replacement parts, and are obsolete as soon as production of the products in which they are used ceases. Once the ASICs for a particular circuit are no longer available from the manufacturer, the repair of equipment that used the ASIC is no longer economically possible.

In recent years the emergence of very complex and powerful field programmable gate array (FPGA) chips has the begun to change the balance back in favour of repair and maintainability. This is because the FPGA chip can be reprogrammed relatively easily, and the parts are generally available from distributors. The use of reprogrammable technology has allowed manufacturers to reduce inventory because it is the software that defines the function of the FPGA and so a single FPGA may be used with different software in several products. Another advantage is that the manufacturer can make enhancements to the product in production without either waiting for new chips or scrapping the ones already in stock. Upgrades can even be performed in the field, often requiring just a laptop and an interface cable.

Despite the large-scale adoption of ASICs and FPGA, a large number of digital chips are still used that are manufactured in huge numbers and from many suppliers. It is important to be able to recognize and replace these devices. The four main features of a chip that is one of a standard family are:

- its type number
- the type of packaging
- the pin numbering and allocation
- the type of output.

Digital logic devices need to be provided with regulated supply voltages. Almost universally now, a single positive regulated supply of $+3.3\,V$ or $+5\,V$ is used, with an earth (zero voltage) return, but a few chips may need both regulated positive and negative supplies and/or different voltages in addition to the earth connection. The supply current can range from a few nanoamps for some types of complementary metal-oxide semiconductor (CMOS) chip to several milliamps for the older transistor–transistor logic (TTL) types. Microprocessors and complex programmable logic devices often use $+3.3\,V$

for their input–output interfaces and a lower voltage, such as 1.8 V, for the core logic. This is mostly to reduce the power consumption, and hence heat dissipation.

All digital devices specify an acceptable range for input and output signal voltage. Because of the development of logic chip types starting with TTL (see later), it has been customary to design other types of digital circuitry to use the same levels (**TTL levels**). These are:

- high level: +2.2 V to +5.5 V
- low level: 0 V to +0.8 V.

Threshold level is another important term that relates to the changeover from level 0 to 1 or from level 1 to 0. If we consider an integrated circuit (IC) that operates with TTL signals whose nominal voltage levels are 0 V and +5 V, the signals that are acceptable at an input need not be exactly at these levels; they will normally conform to the high and low range illustrated above. The values of +0.8 V and +2 V are the threshold levels, and any input between these levels is indeterminate; you cannot be sure what the output will be for an input that is in the threshold range. Digital circuitry must therefore be designed so that these intermediate levels exist only for a very short time during the change from 0 to 1 or 1 to 0.

The **noise margin** of a digital circuit is the amplitude of an unwanted input pulse that will just make the device switch over. This refers mainly to an input at the 0 level. For example, if the maximum possible input voltage level to guarantee logic level 0 is 0.8 V, and the specified maximum output representing a logic 0 is 0.4 V, then a positive-going noise pulse of greater than 0.4 V peak amplitude may be enough to cause the gate to switch over. This value of 0.4 V is the noise margin (Figure 36.1). It is more accurately described as the noise margin for logic 0, but since the noise margin for logic 1 is usually larger, this is the value that is normally quoted.

Noise margins are improved if digital circuits can be operated from higher voltages, so that the typical input levels of 0 V and +15 V that are used by the 4000 series of CMOS chips permit a much larger noise margin. The usual

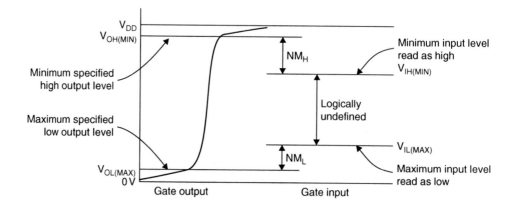

Figure 36.1   Threshold levels and noise margin

TTL type of circuit, however, is restricted to a $+5$ V supply, and noise problems are reduced by using a well-stabilized power supply with decoupling at each chip, and by using low-impedance inputs.

Modern microprocessor and allied chips often use metal oxide semiconductor (MOS) circuitry with a 1.45 V core supply used for processing actions, but using a 3.3 V level for (high) output. This reduces the dissipation within the microprocessor, but maintains a reasonably large noise margin at the output.

The **input current** of a device will be quoted at both signal levels. For TTL families, the input current at level 1 is very low, but the input current for logic 0 is substantial, several milliamps in the older designs. High-speed CMOS logic chips have input currents in the order of $1\,\mu A$, while some special-purpose low-power CMOS logic chips have very low input current figures in the order of $10\,pA$ (10^{-12} A) for either level of input. The output current range is usually between 1 and $5\,mA$ for standard devices, but may be significantly more for high-speed gates or larger special-purpose output buffers.

Rise and fall times are the times needed to change input voltage levels from one threshold voltage level to the other, and are in the order of nanoseconds or picoseconds. A more important figure is **propagation delay**, which is the time between changing the input level and obtaining a corresponding change at the output. These times are in the order of $5\,ns$ for most of the standard families of chips, but times measured in picoseconds (10^{-12} s) are required for some purposes, notably fast ethernet (gigabit ethernet) and optical network components, as well as fast analogue-to-digital (A-D) conversion.

Power dissipation for a chip can be high, several hundred milliwatts, for bipolar types, but for CMOS types it is low, typically several $100\,\mu W$, other than large-scale chips. For many digital circuits, the power dissipation of the standard logic chips is of little importance because so many circuits use fast microprocessor and memory chips dissipating tens of watts or more, so that the cooling requirements are fixed by the needs of these other chips. Nevertheless, a packed logic board that uses only normal standard chip families may dissipate enough heat to require fan-assisted cooling.

The **absolute maximum ratings** for a logic chip define the levels that must not be exceeded, even for short periods. Any breakdown of regulation of a power supply will cause the absolute maximum voltage supply rating to be exceeded, so power supplies must contain some form of **overvoltage cutoff** that will operate in the event of failure of regulation. Another common hazard is excessive input, because an input voltage may be obtained from some external circuit. Although MOS chips are protected by diodes to guard against damage from electrostatic voltages, these protection diodes are not capable of passing the more substantial currents that would pass in the event of a failure of power regulation in a unit that fed signal into a logic circuit.

Two terms that are very important to the operation of digital circuits are **source** and **sink**, applied to the current levels at an input or output of a digital IC. When we say that an output can source $15\,mA$, for example, we mean that the output can provide this maximum amount of current to whatever is connected to it. The connections will be either the inputs of other digital ICs, discrete circuits, or other loads such as micromotors.

The ability to sink current at an output means being able to allow inward current flow without an unacceptable change in voltage level. If we say that the input of an IC can accept a sink of 1 mA, for example, we mean that 1 mA can flow out of the input, and this current will flow to earth through the output terminal of the IC that drives this input, without causing the voltage at that point to rise unacceptably.

This leads to the term **fan-out**. The fan-out of a digital IC means the number of standard gate inputs that can be reliably driven from the output. For example, if a gate output can sink or supply 16 mA, and its input requires sinking of 1.6 mA, then one output can be connected to as many as 10 inputs without compromising the ability of the circuit to operate.

Input connection is important for logic devices. No input must ever be left floating (disconnected) because this can lead to unexpected output changes, particularly at high clock rates when stray coupling can produce input voltages. The old TTL chips could be operated at slow clock rates, with an input left open-circuit, because the input to an emitter automatically ensured that the input level was at logic 1, but later types, and MOS chips in particular, must always have all inputs connected either to the output of another chip, or to one or other logic level through a resistor.

Practical 36.1

Set up the circuit of Figure 36.2, taking care to ensure that the inputs of the unused gates are grounded. Observe the supply current while

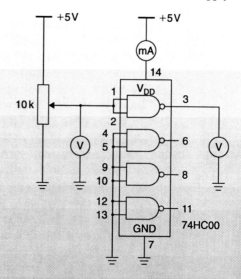

Figure 36.2   Using a 74HC00 CMOS NAND gate to investigate the effect of input voltages other than logic levels on supply current

(Continued)

Practical 36.1 (Continued)

varying the input to the gate from 0 V to 5 V. Plot a graph of the output voltage and supply current against the changing input voltage. Repeat the experiment with a 74LS00 device, if one is available. Comment on the difference in results. What would you check if replacing a faulty 7400 with 74HC00 device in a piece of equipment?

Device lettering and numbering

The two main logic families of devices are TTL and MOS, but there are several important subdivisions in each group. Originally, **TTL** meant a circuit in which each input was to the emitter of an IC transistor, with an output from a collector. A TTL circuit is designed to operate with a +5 V stabilized supply. Any voltage level above about 2.4 V at the input will be taken as being at logic level 1, and any voltage less than about 0.8 V at the input will be taken as being at logic level 0.

For example, Figure 36.3 shows a circuit diagram that approximates to the internal circuitry of a TTL gate of the original 74 family, whose type numbers start at 7400 and continue into five- and six-figure numbers. The input stage consists of a transistor which has been formed with two (or more) emitters. This is comparatively simple to carry out when an IC is being formed, and 15 or more emitters can be formed on a single base of a transistor. The output stage makes use of the familiar totem-pole type of circuit, and the stages between these two will implement the gate action. They need not concern us here.

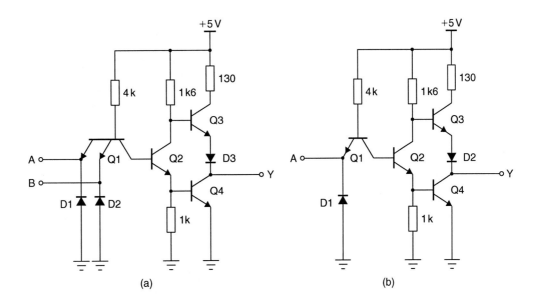

Figure 36.3   Typical circuitry for: (a) standard TTL NAND gate, and (b) inverter

This type of IC family is now termed **standard TTL**, and is almost obsolete. The input impedance is low, so that an input current of about $-1.6\,\text{mA}$ must pass to an input to make the input transistor conduct. The output stage can sink or source a current of up to $16\,\text{mA}$, so that the maximum fan-out is 10.

Bearing these quantities in mind, we can now look at the operation of the typical standard TTL circuit in detail. The collector of the input transistor Q1 (Figure 36.3a) is connected directly to the base of Q2. When both inputs, A and B, are connected to logic 1, always $+5\,\text{V}$, no current will flow between the base and either emitter, since the base voltage is also $+5\,\text{V}$ connected via the 4k resistor. The base-collector junction of Q1 forward-biased current flows into the base of the following transistor Q2, which is thus turned on. In this condition, the top output transistor will be cut off, the base pulled low by the collector of Q2, and the output Y will be low because Q4 is conducting, turned on by Q2 emitter pulling its base up.

Now, if either one of the emitters is connected to logic 0, current will flow through the base of Q1 to the emitter which has been connected to logic 0. Because this current is large enough to saturate the transistor, the collector voltage will be low, and no current will flow into the base of Q2, which will now be cut off. In this condition current flows through the 1k6 resistor into the base of Q3, causing this transistor to conduct and thus pulling the output Y up to the $+5\,\text{V}$ line, via D3. Q4 is cut off, its base tied to $0\,\text{V}$ by the 1k resistor.

Since the output at Y is low only when both inputs are at logic 1, the action is that of a **NAND gate**, and the use of Q4 to connect the output $0\,\text{V}$ ensures that comparatively large currents can pass from the terminal Y to the $0\,\text{V}$ line without raising the voltage level of the output above the guaranteed low-level voltage. Remember that when a transistor is saturated, its collector-emitter voltage is low, typically $0.2\,\text{V}$, and does not rise appreciably when current flows. Ohm's law is not obeyed by a transistor junction because the internal resistance changes as current changes.

Standard TTL chips carry a guarantee that a current of up to $16\,\text{mA}$ can be sunk at the output when Q4 is conducting. Because the current that Q3 can pass is limited by the value of the 130 ohm collector resistor within the IC, the maximum current that can be sourced when Q3 is conducting is limited to about $1\,\text{mA}$.

The input currents flow only from the emitters that are connected to logic level 0, and this is a maximum current of $1.6\,\text{mA}$ per emitter. Since the maximum current (guaranteed) that can be sunk at an output is $16\,\text{mA}$, 10 times the maximum low-level input current for a guaranteed logic 0 level, this fixes the fan-out of a circuit of this type at 10.

All members of the 74 family of standard TTL chips use inputs that are transistor emitters, so that they pass current at logic level 0 and no current at logic level 1. The **propagation delay**, which is the time between changing the level at the input and finding a change at the output, is in the range of 11–$22\,\text{ns}$ ($1\,\text{ns} = 10^{-9}\,\text{s}$). The power dissipated per gate under average switching conditions is $10\,\text{mW}$, and the typical maximum operating frequency is around $35\,\text{MHz}$.

At the time of writing, standard TTL chips are manufactured only for replacement purposes, although hundreds of millions of standard TTL chips are still in use. Another type of TTL circuit, the low-power Schottky transistor–transistor logic (**LSTTL**), largely replaced the standard variety because of the twin advantages of high speed of operation and lower power dissipation. The Schottky chips are identified by the use of LS in the type numbers, so that type 7400 is an standard TTL NAND gate, but 74LS00 is a low-power Schottky NAND gate. A faster version of Schottky clamped TTL is also available, designated 74S00, for example.

Low propagation delays that are obtained in standard TTL by using large currents inevitably lead to larger chip dissipation, because the current is flowing through the integrated components which are all part of the IC. The LS range of TTL ICs avoids this difficulty by using a different principle, relying on the use of Schottky diodes, components that can be made easily in IC form.

The Schottky diode, whose symbol is illustrated in Figure 36.4(a), is formed by evaporating aluminium onto silicon, and its remarkable feature is its very low forward voltage when it is conducting, in the order of 0.3 V. This feature is used in two ways. One application is to carry out the logic action using the diodes in circuits similar to those used in the early diode transistor logic (DTL) ICs. The other use is in preventing the transistors in the circuit from saturating. A transistor is saturated when it is fully conducting, with the

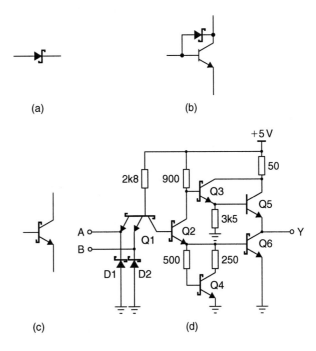

Figure 36.4 (a) Schottky diode symbol, (b) arrangement of Schottky diode and transistor, (c) Schottky transistor symbol, and (d) Schottky clamped TTL NAND gate

base passing more current than is needed to make the collector circuit conduct fully, since the collector current is limited by the value of collector load resistor. In this state, the collector-to-emitter voltage will be low, about 0.2 V, compared with the 0.6 V which will exist between the base and the collector.

When a transistor is saturated, there is a comparatively large amount of slow-moving charge in the base layer, and when the transistor is switched off by connecting the base terminal to the voltage level of the emitter, this stored charge will permit current to flow between the collector and the emitter for a time which can be as long as a microsecond, until the charge is neutralized. This restricts the speed at which a switching circuit can be operated, but if the transistors in switching circuits are not allowed to saturate, a very considerable increase in switching speed is possible. This cannot, however, be achieved by normal biasing methods, particularly within an IC. Another advantage of using Schottky junctions is that they operate using majority carriers, so that the storage effect does not apply.

Figure 36.4(d) shows the typical circuit of a NAND gate of the 74S family which uses Schottky transistors for logic, and also Schottky diodes for the input protection (D1 and D2)

The Schottky connection within the transistor is between the base and the collector (Figure 36.4b), so that when the transistor collector voltage approaches the saturation value, current from the base circuit is diverted by the Schottky diode into the collector circuit. This prevents the base current from reaching the value that would cause saturation. The Schottky symbol is shown on the transistor base (Figure 36.4c) to make it clear that the Schottky diode structure exists within the transistor. The transistor circuit that is used within these Schottky TTL ICs is designed to make use of current stabilizers in addition to other methods of preventing saturation. Schottky clamped TTL gates designated 74S are much faster than their standard TTL counterparts, but this comes at the expense of increased power dissipation; for this reason, a low-power version was introduced, the 74LS series.

The 74LS series operates at the same voltage levels as the standard TTL 74 series, but the maximum output current is 8 mA and the input current at logic level 0 is −0.4 mA, one-quarter of the standard TTL current, so that the fan-out is 20. The propagation delay is in the range 9–15 ns, appreciably lower than that of standard TTL, and the average power per gate is about 2 mW. The typical switching frequency is 40 MHz.

Although most of the gates of the 74, 74S and 74LS series of TTL digital ICs make use of the **totem-pole** type of output circuit which has been illustrated, a few gates use **open-collector outputs**. This means that the phase-splitter and one-half of the output stage is missing, so that the output terminal of such a gate is simply the collector terminal of a transistor. If this gate is to be used in normal logic circuit applications, a load resistor, connected between the output and the +5 V supply line, must be added externally. Such outputs are used only in a few circuits where the outputs of several gates are connected, and the open-collector construction prevents a burnout when one transistor is at a high output and another connected to it is at a low output.

CMOS gate circuits

The most common semiconductor technology in current use is that of metal oxide semiconductor field-effect transistors (MOSFETs). Three types of IC can be manufactured using P- and N-channel field-effect transistors (FETs). **PMOS** ICs use P-channel FETs exclusively, **NMOS** ICs use N-channel FETs exclusively, and **CMOS** (C meaning complementary) ICs make use of both P- and N-channel FETs in a single circuit. PMOS methods were initially used for manufacturing microprocessors and similar chips, but were superseded by NMOS. Fast versions of CMOS are used in laptop computers.

One family of CMOS devices uses the 4xxx type of number, illustrated later, but the most common types currently in use follow the 74xx type of numbering, with lettering to distinguish the types, such as 74HC, 74HCT and 74AHC.

A typical CMOS NAND circuit, like that of one gate of the CD4011A or 74HC00 quad NAND IC is shown in Figure 36.5(b). In this circuit, M1 and M2 are both P-channel types, whereas M3 and M4 are N-channel types. The P-channel FETs will be switched into conduction by a logic 0 input at their gates, since their sources are connected to the positive supply.

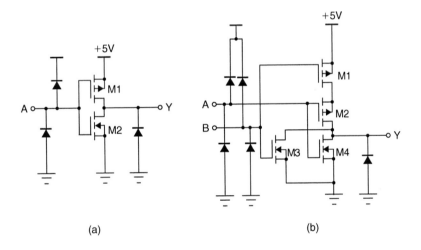

(a)   (b)

Figure 36.5 (a) Typical CMOS inverter, and (b) NAND gate

The N-channel FETs will be switched into conduction by a logic 1 input at their gates, because their sources are connected to the 0 voltage line. With either or both gate inputs at logic 0, the P-channel FETs will conduct, keeping the output high. Only when both inputs are high can both N-channel FETs conduct, and thus connect the output to the logic 0 level. The action is therefore that of a NAND gate. A NOR gate can be created using the same set of components by connecting the N-channel FETs in parallel and the P-channel FETs in series.

The 4000 series CMOS ICs can operate with a wide range of supply voltages, typically 3–15 V, and with very small currents flowing, 5 μA or less. The logic 0 and 1 voltages are normally very much closer to the supply voltage levels than is possible with older bipolar designs of the TTL, ECL or I2L types. For example, using a +5 V supply, a logic 1 voltage of +4.95 V and a logic 0 voltage of 0.05 V can be obtained. This makes for much better noise margins than can be achieved with either standard TTL or LSTTL devices.

The input current is always negligibly small because the inputs are connected to MOSFET gates, and the output currents of unbuffered gates are typically about 0.5 mA maximum. The fan-out figure for low-frequency operations can be very large, 100 or more, but the value decreases as the frequency of operation is increased. This is because the small currents that are available at the output must be capable of charging and discharging the capacitance at the input of each gate that is connected to the output. This requirement for charging and discharging stray capacitances also increases the total dissipation of the IC as the frequency of operation is increased.

A simple gate, for example, which has a dissipation of 1 μW at a frequency of switching of 1 kHz, may have a dissipation of 0.1 mW at an operating frequency of 1 MHz. This factor limits the operating speed of the earlier types of CMOS circuit, and leads to these types (the 4000 series) being used in low-speed applications rather than for high-speed machine control or computing applications. They are widely used where speed is not of primary importance.

The very high insulation resistance of the gates makes them very susceptible to damage from electrostatic charges, and modern CMOS ICs are manufactured with a network of diodes connected to the inputs and outputs (Figure 36.5), which will conduct whenever the voltage between gate and source or gate and drain becomes excessive. These diodes will typically protect a device for static voltages of up to 4 kV, but if higher voltages are likely to be encountered stringent earthing precautions must be taken. For example, operators may be required to use metal wrist straps that are earthed, and work on a conducting earthed surface. The safest way to work with CMOS devices is to earth all pins together until they are inserted into place and connected.

Note that walking along a nylon carpet can generate voltage levels in excess of 16 kV.

The original 4000 CMOS family of ICs typically used a 7 μm metal gate CMOS process which gave an absolute maximum supply voltage of about 18 V, although in operation the supply should be kept below about 15 V. Numbered from 4000 upwards, most of the standard functions were in the 40XX range, with some more complex ICs in the 45XX range. The numbers are suffixed with the letters B and UB; these suffixes indicate buffered and unbuffered devices, respectively. Buffered CMOS devices use a larger output stage to buffer the logic function, providing a higher current output. Since each stage is an inverter, a buffered device uses a chain of two inverter stages following the logic gate, and this adds to the propagation delay of the device.

The newer families of CMOS, based on a silicon gate process, used the numbering of TTL devices, with the letters HC, AC or LV, etc., so that a 74HC00 is a CMOS gate equivalent to the 7400 or 74LS00 in action, but with greatly reduced power consumption. These CMOS equivalents can be used, subject to some caution on their characteristics, as replacements for older bipolar standard TTL and LSTTL devices.

The older 4000B series have typical propagation delays of 125–250 ns, power dissipation per gate of 0.6 μW and a typical switching frequency of 5 MHz. The unbuffered UB family features propagation delays of 90–180 ns and slightly higher typical frequency ratings. The 74HC family comprises direct replacements for 74LS types, with propagation times of 8–15 ns, power dissipation of 1 μW and typical frequency of 40 MHz. The 74HCT is very similar, but with slightly longer propagation delays; the T indicates that the input thresholds are set to match TTL levels rather than standard CMOS levels, i.e. they are fixed voltages rather than being a proportion of the supply voltage. One of the more recent CMOS families is designated 74AC, with propagation delays of, typically, 5 ns, power dissipation of 1 μW and a typical frequency of 100 MHz. The 74LV family is optimized for very low supply voltages in the range 1.0–3.6 V; these devices are often used as 'translators' between low-voltage microprocessors, e.g. 2.7 V, and standard 5 V logic systems.

The lettering used for more recent 74-series logic chip families can be confusing, although the most common types have already been listed. Table 36.1 shows a more comprehensive list of known types in the year 2006. The quest for higher speed and lower dissipation continues, however, and several of the types in this list may become obsolete, with new types being introduced, during the lifetime of this book.

Table 36.1   Some 74 series logic families commonly available in 2006

Prefix	Description	Prefix	Description
74ABT	Advanced bi-CMOS technology	74BCT	Bi-CMOS bus interface technology
74AC(T)	Advanced CMOS (TTL input levels)	74F	Fast TTL
74AHC(T)	Advanced high-speed CMOS (TTL input levels)	74GTL	Gunning transceiver logic
74ALB	Advanced low-voltage bi-CMOS	74HC(T)	High-speed CMOS (TTL input levels)
74ALS	Advanced low-power Schottky logic	74HCU	High-speed CMOS unbuffered
74ALBT	Advanced low-voltage bi-CMOS	74LS	Low-power Schottky
74ALVC	Advanced low-voltage CMOS	74LV	Low-voltage CMOS
74ALVT	Advanced low-voltage bi-CMOS	74LVC	Low-voltage CMOS (5 V tolerant inputs)
74AS	Advanced Schottky logic	74LVT	Low-voltage bi-CMOS
74AVC	Advanced very low-voltage CMOS	74LVTZ	Low-voltage bi-CMOS live insertion
74AVHC	Advanced very low-voltage high-speed CMOS	74S	Schottky clamped TTL

Table 36.2 shows a comparison of the most common families of devices that use either the 7400 or the 4000 type numbers.

Some circuits, notably computer boards, feature chips that are described as having **tri-state outputs**. This does not mean that they use three logic states, only that the chip can be isolated from inputs and output by not applying an enabling pulse. The purpose of this is to allow a set of chips to be permanently connected to the same set of input and output lines, selecting which chips are in use by their enabling inputs. Tri-state logic chips are

Table 36.2   Summary of logic family characteristics

	TTL	LSTTL	4000 CMOS	74HC	74AC
V+ supply	5 V	5 V	3–15 V	2–5 V	2–5 V
$I_{max/1}$	40 µA	20 µA	<0.2 µA	<1 µA	<1 µA
$I_{max/0}$	−1.6 mA	−0.4 mA	<0.2 µA	<1 µA	<1 µA
$I_{max/out}$	16 mA	8 mA	1.6 mA[a]	8 mA	15 mA
Delay	11–22 ns	9–15 ns	40–250 ns[a]	10 ns[a]	5 ns
Power	10 mW	2 mW	0.6 µW[a]	<30 µW	<30 µW
Frequency	35 MHz	40 MHz	5 MHz	40 MHz	100 MHz

V+ supply: normal positive supply voltage level; $I_{max/1}$: maximum input current for logic level 1; $I_{max/0}$: maximum input current for logic level 0; $I_{max/out}$: maximum output current; Delay: propagation delay in nanoseconds (a low delay means a fast device); Power: no-signal power dissipation per gate in mW or µW; Frequency: typical operating frequency.
[a] These quantities depend on the supply voltage level.

therefore extensively used in any circuits that make use of bus lines to connect different portions of the circuitry.

The packaging of the standard families of logic gate circuits is usually either plastic dual in line (DIL), for pin through hole parts, or small outline (SO) surface mount package. Leadless ceramic chip carrier (LCC) packages are available for high-reliability applications.

Gates are usually packaged in groups of four or six in a 14- or 16-pin package. Single gate chips are also available in surface-mount packages, either SOT23 or SC70; for example, the 74HC1GU04 is a single inverter in a SOT23 package. Such devices are commonly used in mobile devices to modify or buffer the output from an ASIC, etc.

The standard 14- and 16-pin logic ICs have two pins reserved for earth (pin 7 or 8) and V+ (pin 14 or 16). The conventional way to show pin allocations is a diagram that shows the top view of the chip with logic gate symbols and connections drawn in as illustrated in Figure 36.6.

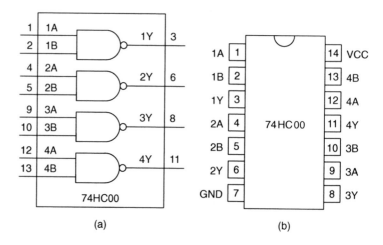

Figure 36.6   Typical pin-out diagram for a logic chip

Multiple-choice revision questions

36.1 How is the noise margin of a logic family defined?
 (a) $V_{OH} - V_{OL}$
 (b) greater of $V_{DD} - V_{OH}$ and $V_{OL} - GND$
 (c) smaller of $V_{IL} - V_{OL}$ and $V_{OH} - V_{IH}$
 (d) $V_{IH} - V_{IL}$.

36.2 A logic family has an input current of 1.6 mA and an output current of 8 mA. What is the fan-out of the family?
 (a) 8
 (b) 5
 (c) 20
 (d) 10.

36.3 What parameter causes the main limit on fan-out of CMOS logic in high-speed applications?
 (a) d.c. input current
 (b) output current
 (c) input capacitance
 (d) power supply voltage.

36.4 How are CMOS circuit inputs protected from electrostatic discharge damage?
 (a) bipolar input stages
 (b) low power supply voltage
 (c) using both NMOS and PMOS transistors
 (d) input protection diodes.

36.5 Which common logic family can be used over a 3–15 V supply range?
 (a) 74LS TTL
 (b) 4000 CMOS
 (c) 74HC CMOS
 (d) standard TTL.

37 Sequential logic

Sequential logic circuits are those whose outputs depend not only on the current inputs, but also on the order that the inputs change. This means that the output depends on past as well as present inputs, so these circuits have 'memory' and are a very important building block for many complex systems, such as counters, timers and computers.

A **flip-flop** is a circuit of the bistable type. Integrated circuit (IC) flip-flops tend to be of complex designs that would be uneconomical to manufacture with discrete components. The simplest flip-flop is the R-S type.

The simple R-S flip-flop changes state when either of the inputs becomes, or is set to, zero. Most digital circuits require flip-flops that change state only when forced to do so by means of a pulse, called a **clock pulse**, arriving at a separate input. In a set of such **clocked flip-flops**, all the flip-flops in a circuit can be made to change state at the same time.

Clocking has the advantage that so long as the inputs can be guaranteed stable by the time the clock pulse arrives, the order that the inputs change in, or changes of state at the inputs at any other time, have no effect on the output. Between clock pulses, the output remains as it was when set by the last clock pulse.

The edge-triggered D-type flip-flop has inputs of data (D), clock (CLK), preset (PRE) and clear (PRE); with outputs of Q (main output) and Q# or $\overline{Q}$, the inverse of the Q output. This flip-flop is triggered on the **leading edge** of the clock pulse. The logic value on the D (data) input is then transferred to the Q output, thus delaying the data by the period of one clock pulse. The truth table is illustrated in Table 37.1.

Table 37.1	D-type flip-flop truth table				
$\overline{PRE}$	$\overline{CLR}$	CLK	D	Q	$\overline{Q}$
0	1	X	X	1	0
1	0	X	X	0	1
0*	0*	X*	X*	1*	1*
1	1	↑	1	1	0
1	1	↑	0	0	1
1	1	0	X	Q_0	$\overline{Q}$

X: either state; ↑: goes high; *: $\overline{PRE}$ and $\overline{CLR}$ both zero, prohibited output undefined.

Reference to the truth table shows the action of the preset and clear inputs to set the initial output states of the device. When (PRE) = 0, Q is set to 1 and when $\overline{CLR}$ = 0, Q is set to 0. The states of both CLK and D are irrelevant (X) at this moment in time because (PRE) and $\overline{CLR}$ override them. Only when $\overline{\text{PRE}} = \overline{\text{CLR}} = 1$ will the clock pulse transfer the data, and even then only as the clock pulse reaches a high value.

The state where $\overline{\text{PRE}} = \overline{\text{CLR}} = 0$ represents an unstable condition in a D-type flip-flop, and is prohibited.

The **D-type latch** (Figure 37.1), sometimes called a **transparent latch**, is very similar to a D-type flip-flop, except it is level rather than edge triggered; that is, the outputs follow the inputs while the CLK pin is high and are held or latched at the last state once the clock pin goes low. When **latched**, the output value at the Q terminal remains at its previous setting, regardless of the input at D, until a clock pulse is applied. A latch is a very valuable component of digital circuits because it performs a form of memory action. The output ports of microcontrollers, for example, use a latch to hold the output data until it needs to be changed by the program.

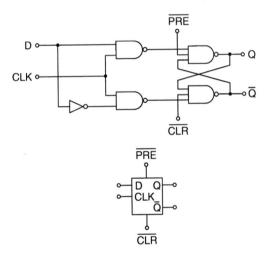

Figure 37.1   Schematic and symbol for D-type latch

The **J-K flip-flop**, which also contains steering logic to avoid the output problem of the R-S flip-flop at R = 0, S = 0, also has a greater range of output controls, as shown by Table 37.2. This type of flip-flop has three signal inputs labelled **S** (set) and **R** (reset) (or sometimes preset and clear) which work independently of the clock pulses, and the usual two outputs Q and Q# (with Q# always the inverse of Q for any combination of inputs). The **J** and **K** inputs are programming inputs that determine what will happen when a clock pulse is applied.

If the complementary outputs of the J-K device change state before the end of a clock pulse, then because of the internal feedback, the inputs will also change. This can cause the device to oscillate until the end of the pulse and leave the output in an indeterminate state.

Table 37.2   State table for a J-K flip-flop

J	K	Q_0	Q	Note
0	0	0	0	No change
0	0	1	1	No change
0	1	0	0	Clear
0	1	1	0	Clear
1	0	0	1	Set
1	0	1	1	Set
1	1	0	1	Toggle
1	1	1	0	Toggle

Q_0: state now; Q: state after next clock pulse.

To avoid this **race condition**, the clock pulse duration should be small compared with the propagation delay. For high-speed operation, this is avoided by using a **master–slave** device as shown in Figure 37.2, where the NAND logic gates act as switches. The state table is the same as for the simple J-K device.

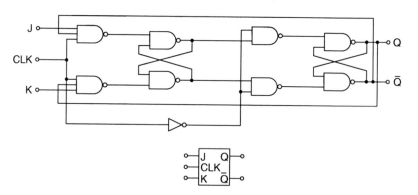

Figure 37.2   Master–slave J-K flip-flop and symbol

Data is transferred in two stages on *both* edges of the clock pulse. On the positive edge, the input gates are enabled and allow the data to be loaded into the master flip-flop. At this time the output logic gates are open-circuit. On the negative pulse edge, the switch states reverse, the master is isolated from the input and the data is transferred to the slave stage to provide the output. The master–slave design is preferred in most applications.

The binary counting action occurs when both J and K inputs are taken to logic 1 [+5 V in the case of transistor–transistor logic (TTL) circuits]. Two clock pulses arriving at the clock terminal give rise to one pulse at the Q output, while a pulse of inverse polarity will simultaneously reach the Q# output.

• Clock pulse inputs to digital circuits must have short rise and fall times, and the inputs to gates should never be slow-changing waveforms. The reason for this is that the gain of these circuits, considered as amplifiers, is very large, so that a slow-changing waveform at the inputs can momentarily bias the circuit in a linear mode, allowing it to amplify noise signals present in the circuit, which appear as a burst of pulses or oscillations. Oscillation can cause multiple counts and erratic operation.

- Logic circuits designed to tolerate slowly changing inputs use a Schmitt trigger input stage which uses hysteresis to reinforce the switching action and reduce the possibility of oscillation.

Practical 37.1

Connect the circuit shown in Figure 37.3 using a 74HC74 IC on a breadboard. The 74HC74 contains two D-type flip-flops. Apply clock pulses either from a slow pulse generator or from a debounced switch (see Chapter 15), and observe the output indicator light-emitting diodes (LEDs). Now connect the clock terminal (pin 11 of the second flip-flop in the package) to the Q output (pin 5 of the first flip-flop), and reapply the slow clock pulses. Observe the indicators. Remember that no input pin may ever be left unconnected.

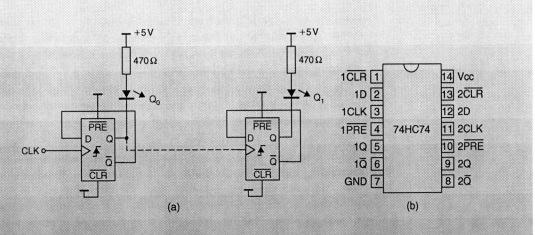

Figure 37.3 Circuit for practical: (a) connections, and (b) pinout of 74HC74 IC

A fundamental feature of all clocked bistable devices is that they can be used to divide the input number of clock pulses by two. Thus, for every two clock-pulse inputs, the Q output changes once. This concept forms the basis of electronic binary counters, an example of which is shown in Figure 37.4.

The pulse sequence to be counted is the input to FF_1, whose Q output now provides the clock for the next stage, and so on. This is an **asynchronous counter**. A count input to the first stage thus ripples through the circuit from one flip-flop to the next, which is why a counter of this type is often called a **ripple** or a **ripple-through** counter. Note that the output

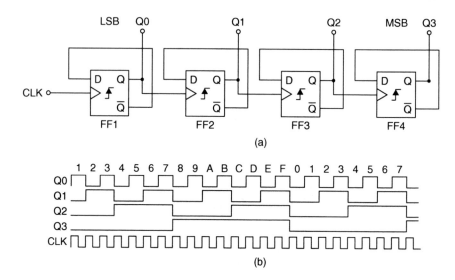

Figure 37.4   Ripple or asynchronous counter: (a) circuit, and (b) counter waveforms

count should be read from right to left, because the most significant bit (**MSB**) is, by convention, shown on the right of the drawing. A chain of n flip-flops will, connected in this way, produce a counter that divides the number of input pulses by a maximum of 2^n.

The ripple counter waveforms are shown in Figure 37.4(b). Since the $\overline{Q}$ output is fed back to the D input, the Q output toggles, or changes, on the rising edge of every clock cycle.

Ripple counters are unsuitable for applications such as timers where their output value needs to be decoded, because the delays incurred by the ripple progressing through the counter mean that the parallel decode output can have false-positive states. Although these are often only very short lived, of the order of tens of nanoseconds, the result can be to trigger erroneously the decoding logic. For this reason synchronous counters are always used where decoding of the count is required. Ripple counters are useful for frequency division, as in clock scaling and similar applications.

If the design of the counter is changed so that all the flip-flops are clocked at the same time (**synchronously**), the propagation delay is reduced to that of a single flip-flop, thus eliminating the race hazard and making it possible to decode the count successfully. For TTL counters, the maximum count frequency is about 50 MHz, compared with about 25 MHz for a 4000 CMOS counter. 74HC high-speed CMOS counters can operate at much higher speeds.

Synchronous counters can be constructed using J-K flip-flops and gates, and a typical circuit for a synchronous 4-bit counter is shown in Figure 37.5(a). The J-K flip-flop changes state (toggles) only when J = K = 1 at the time of a clock pulse. The toggle action in this circuit is controlled via the Q outputs and the AND gates. Before any bistable can toggle, all the earlier

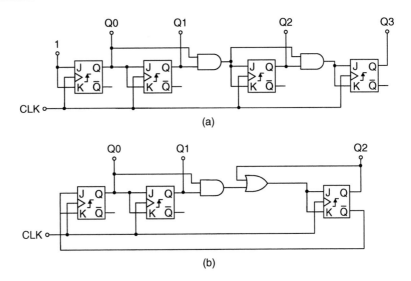

Figure 37.5 (a) Synchronous 4-bit binary counter, and (b) synchronous divide by 5 counter

bistable Qs must be at logic 1. Because J and K inputs of the least significant bit (LSB) flip-flop are connected to logic 1 it toggles on every clock pulse. The second only toggles while the Q output of the first is high, the third while both the first and second flip-flops have high Q outputs, and so on.

You may need to divide by a number N which is not a power of 2; for example, to divide by 5 for part of a denary counter (Figure 37.5b). A counter that carries out a division by N is called a **modulo-N counter**. There are several methods of designing such a counter. The **state machine** method defines the valid counts or states and then logic is designed to achieve the correct progression from each state to the next. Figure 37.5(b) shows a divide by 5 circuit that is made in this way. Another design method that can be used for such counters is that of an **interrupted count**, meaning that the counter contains sufficient flip-flips for a longer count, but is interrupted after the count of N and reset to zero. This requires decoding logic for the value of N. Another method is the dual modulus counter. This is a counter that divides by N for part of the time and by M for the rest of the time. This allows average fractional counts and is often used in PLL frequency synthesizers.

To construct an interrupted count counter n flip-flops are needed, such that n is the smallest number for which $2n > N$. A decoder circuit detects when a count of N has been reached so that the circuit can be reset to zero. For a decimal counter $n = 4$, because $2^4 = 16$, which is the nearest number greater than 10. Denary 10 is 1010 in binary, so the feedback must detect this pattern to reset the counter to zero.

Integrated circuit counters

A count from zero to nine, followed by resetting, however it is achieved, is the basis of **BCD counters**, in which each counter unit (a set of four flip-flops) is used for one column of a denary number. BCD counting allows for the display of a count number to be simply decoded in denary rather than in binary, so BCD counters are preferred for any application in which a count is to be displayed rather than simply used.

BCD counters are available in IC form in the 74 series, and the simplest unit is the 7490. This is one of the very few IC counters that is classed as asynchronous, and is designed mainly for use with displays. The design is a split counter with one scale of two (a single flip-flop) and one scale of five which uses three flip-flops connected as a synchronous counter. For decade counting, the output of one unit is connected to the clock input of the other, so that the overall action is asynchronous. Because of the asynchronous action, the Q outputs will not change simultaneously, and this can give rise to spikes on the outputs. These are unimportant if the output is used to operate a display, since the display does not store the spike output, but it makes a counter of this type unsuitable if the outputs are connected to latches.

The pinout for the 7490 is shown in Figure 37.6(a). The pins labelled MR_1 and MR_2 are reset inputs; which must both be taken to logic 1 to reset the counter, since they are inputs to a NAND gate. The inputs MS_1 and MS_2 are similarly used to set the most significant and least significant flip-flops to 1 (setting the count at 1001, denary 9). Table 37.3 shows how the MR and MS inputs control the action.

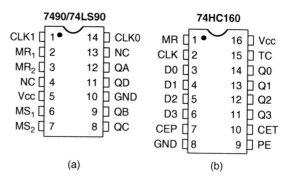

Figure 37.6   (a) Pinout for 7490, and (b) pinout for 74HC160

Table 37.3	Use table for the 7490 type of ripple counter chip						
MR_1	MR_2	MS_1	MS_2	Q_0	Q_1	Q_2	Q_3
1	1	0	X	0	0	0	0
1	1	X	0	0	0	0	0
X	X	1	1	1	0	0	1
0	X	0	X	Counting			
X	0	X	0	Counting			
0	X	X	0	Counting			
X	0	0	X	Counting			

The counter is clocked by the **trailing edge** of the clock pulse. The maximum ripple delay, from the clock pulse input to the Q3 output, is in the order of 100 ns for standard TTL, but around 50 ns for the LS version of the chip.

For decade counting with synchronous counters we would normally make use of an integrated unit such as the 74HC160 type, whose pinout is shown in Figure 37.6(b). This is not so simple as the 7490, considered earlier. To start with, this is a totally synchronous counter, and its internal flip-flops are triggered at the leading edge of the clock, with the outputs also changing at this time. The counter is fully presettable, meaning that the count number can be preset at any stage so that a count from, for example, 3 to 9 rather than 0 to 9 can be carried out if necessary. Only an up-count is available, and the counter allows a hold state in addition to its presetting state, making it fully programmable.

The MR input is a reset input that is active when taken low. Using an RC network to delay the rise of the MR input at power-up can reset the counter, before counting begins. This is often necessary because in any device that consists of a set of flip-flops, application of power will cause some flip-flops to set to a 1 output, while others to reset to a 0 output. In some applications a counter cannot therefore be used immediately power has been applied; it must be reset before any counting pulses are applied. The reset action, as usual, is completely asynchronous and this is reflected in the table by the X, which means 1 or 0 (the don't care state). The CP pin takes the clock pulse input, and the arrows in the table are a reminder that the changes take place on the leading (0 to 1) transition of the clock pulse, which should have a short rise time.

The CEP and CET inputs are normally high for counting, but taking either or both to level 0 will cause the count output to hold its existing state. This action must not be used while the clock pulse input is low, only after the leading edge of the clock pulse when the clock input has settled to a high state. The PE pin is used for parallel loading of the flip-flops, allowing a number to be preset before counting starts or restarts. Figure 37.7 shows a typical sequence of reset, load, count and hold (count inhibit) actions.

Parallel loading is enabled when the PE pin voltage is taken to 0, and in this state each Q output will take the state of its corresponding D input. Figure 37.8 shows a five-decade counter (to a count of 99999) which makes use of these units, connecting them together so that synchronous counting is preserved. Note that synchronous decade counters can be connected to each other in an asynchronous way, with the output of a counter driving the clock of the next, but the circuit shown here preserves fully synchronous action, using one counter to gate the following counters.

Suppose that we want to count in scales other than 8-4-2-1 binary or BCD. The gating methods that are used for synchronous counters allow for any counting system to be used, and design of the **state machine** is carried out making use of state tables to show what J and K entries are required, and diagrams called **Karnaugh maps** to show what gating can achieve this. Karnaugh maps allow the designer to evaluate graphically the simplest representation of a function; however, in practice they have been largely superseded by computer-based design tools.

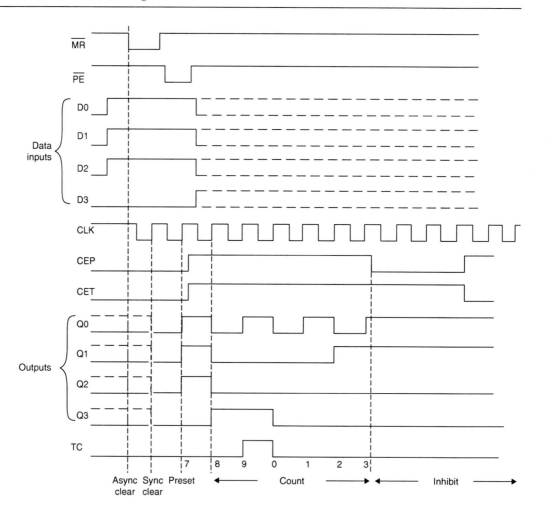

Figure 37.7 Typical operating sequence of the 74160 counter, reset counter, preset with 7, count, inhibit

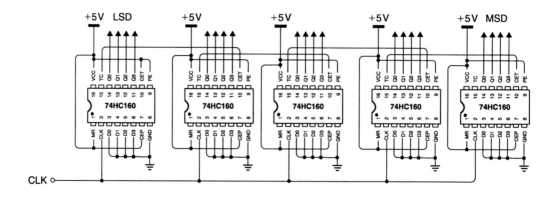

Figure 37.8 Five-decade counter using 74HC160 chips, drawn as it might be built

A binary count can be achieved to any count number by repeating a basic design. This is not necessarily true of other types; using Karnaugh maps, a practical limit of four stages is reasonable. Typically, complex counting systems would be implemented in a microcontroller, application-specific integrated circuit (ASIC) or field programmable gate array (FPGA), and design software would be used to determine the required states and their implementation.

Practical 37.2

Assemble and test modulo-n (where *n* is any whole number) counters, both synchronous and asynchronous. Use ICs such as 74LS74, 74LS112, 74LS193 and 74LS390.

Shift registers

A register is formed by interconnecting a number of bistables in a similar way to a counter. Each bistable forms a memory cell for one bit of information. The data may be input in serial or parallel form and read out in a similar manner, depending on the method of interconnection. There are thus four basic forms of shift register:

- serial in, serial out (SISO)
- serial in, parallel out (SIPO)
- parallel in, parallel out (PIPO)
- parallel in, serial out (PISO).

Some IC implementations are shift registers of one form only, while others are more universal, with the mode of data transfers controlled by the logic values of separate control lines. The general principle is explained using Figure 37.9, where a simple basic register is formed from D-type flip-flops (R-S flip-flops may also be used). Remember that the logic value on the D input is transferred to the Q output on the next clock pulse. In this example, each data value input thus ripples through the register in a manner similar to the counter.

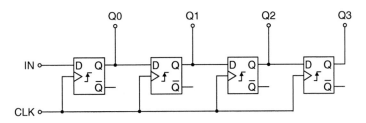

Figure 37.9   Simple 4-bit shift register

The functions performed by shift registers include:

- multiplication and division (SISO)
- delay line (SISO)
- temporary data store or buffer (PIPO)
- data conversion; serial to parallel and vice versa (SIPO and PISO)
- serial input–parallel output shift register.

Figure 37.10 shows the simplified internal schematic of the 74HC194, a universal version of a 4-bit shift register based on D-type flip-flops with AND-OR control logic. This provides for synchronous parallel data loading with serial and parallel output, and control of shift direction.

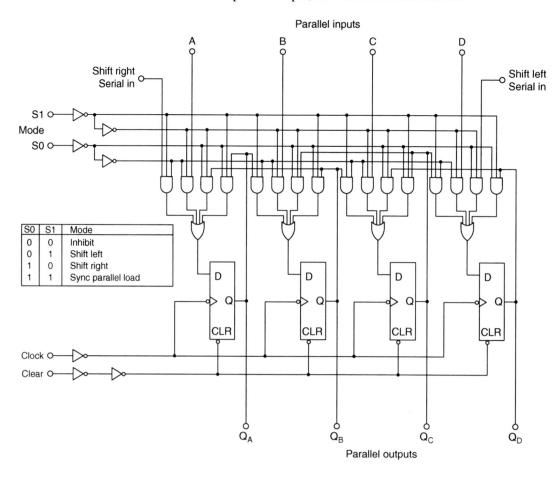

Figure 37.10   Simplified schematic of the 74HC194 4-bit universal shift register

The function of shift register is controlled by two mode select inputs S0 and S1, which determine the direction of shift and synchronous parallel load, or inhibit the shift register when both lines are low. The mode inputs drive a 4 to 1 multiplexer in each stage, consisting of four three-input AND gates and a four-input OR gate. This allows the state of the mode lines S0 and S1 to determine where the D input bit of the flip-flop comes from. With the S0 line set to logic 1 and the S1 line set to logic 0 the leftmost AND gate of each group is enabled. This carries the shift right bit to each stage. Setting S0 and S1 both high selects the next AND gate, which enables the parallel input, so that on the next clock pulse the parallel input is synchronously latched. Similarly, when S0 is set low and S1 set high the third AND gate from the left of each group is enabled and the shift left bit passed to the D flip-flop. When both S0 and S1 are low, the rightmost AND gate is enabled and the output of the stage is fed back to its input, meaning that clock pulses do not cause any change in the register state.

Practical 37.3

Using ICs such as the 74HC194 (Figure 37.11), assemble and test shift registers of each basic type. Once you have tried the various operations, use two shift registers to transfer parallel data via a serial link. How many connections do you need between the two shift registers?

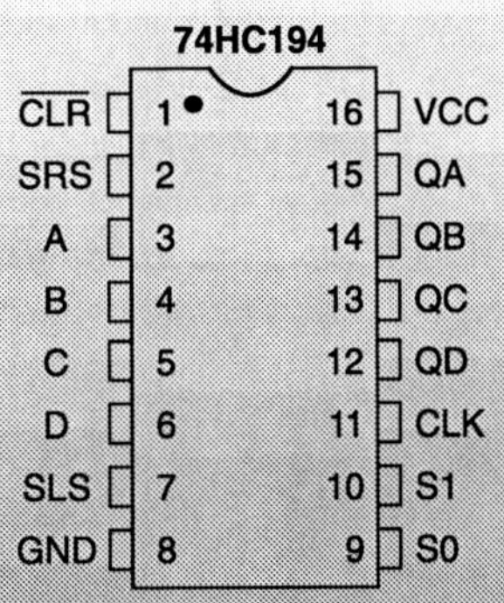

```
              74HC194
    CLR [ 1 •        16 ] VCC
    SRS [ 2          15 ] QA
      A [ 3          14 ] QB
      B [ 4          13 ] QC
      C [ 5          12 ] QD
      D [ 6          11 ] CLK
    SLS [ 7          10 ] S1
    GND [ 8           9 ] S0
```

Figure 37.11   Pinout of the 74HC194 4-bit universal shift register

Ring counters

A shift register shifts the bits one place on each clock, so if a right shifting shift register is loaded with 1000 and clocked three times the result will be 0001. This assumes that it loads zeros to the LSB from the left. This can be viewed as counting in powers of 2; that is, it was loaded with 2^0 and after the first clock the output was 2^1, then 2^2 and finally 2^3 after three clocks. If the output of the last stage is fed back to the input of the first one, as illustrated in Figure 37.12, it will cycle through these powers of two continuously.

Figure 37.12 shows a set of four flip-flops connected in the usual Q to J, $\overline{\text{Q}}$ to K shift register mode, with a reset arrangement that resets all but the LSB flip-flop, setting a logic 1 into the LSB. After resetting, each clock pulse will then circulate that 1 by one place to the right and so back to

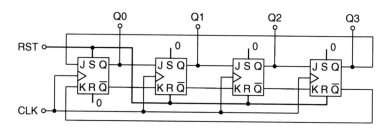

Figure 37.12 Straight-ring counter circuit using a SIPO register. The register must be set up with one 1 bit, since it cannot count if all outputs are zero

the first flip-flop again. Such a counter can be very useful for sequencing because the other method of using a conventional binary counter and decoding logic requires many more gates to implement.

Ring counters can be set up to count arbitrary integer numbers of clock pulses. Simply by setting the number of stages, a seven-stage ring can be used to divide the clock signal by 7, giving a single high pulse out once for every seven clock pulses. Such counters can also be very much faster than conventional synchronous counters.

A disadvantage of the ring counter for dividing is that the output pulse is typically nothing like 50% duty cycle. One way around this problem, which also halves the number of flip-flops required, is to use a switch-tailed ring counter. As the name suggests, this is simply a ring counter with an inversion in the feedback (Figure 37.13).

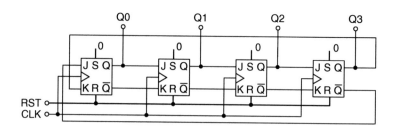

Figure 37.13 Switch-tailed ring counter

The switch-tailed ring counter, also called the Johnson counter or twisted ring counter, has another advantage, in that the reset circuit is simpler: resetting all the flip-flops is sufficient. Figure 37.13 shows a four-stage counter, hence it is a mod-8 counter, the output going through one cycle for every eight input clocks.

The count of a Johnson counter is $2 \times n$, where n is the number of stages, rather than the 2^n that can be achieved with true binary (8-4-2-1 or Gray) counters. A four-stage Johnson counter will therefore give a scale of nine rather than the scale of 16 that a true binary counter can provide. The Johnson numbers always consist of a block of 1s and a block of 0s with no alternation of digits.

Ring counters of the type shown so far have a disadvantage that if, for some reason, whether a power glitch or other transient effect, they reach a number not in their designed sequence, they will count with the new pattern without returning to the intended sequence, until reset. To avoid this lockout the feedback signal can be taken from the last two stages using an AND gate. This has the effect of removing patterns such as 101 from the sequence until the original sequence is recovered. A Johnson count for a four-stage counter is shown in Table 37.4 for the shift-register connections illustrated in Figure 37.13.

Table 37.4	Count table for four-stage Johnson counter				
Clock pulse	Q_0	Q_1	Q_2	Q_3	$\overline{Q}_3$
0	0	0	0	0	1
1	1	0	0	0	1
2	1	1	0	0	1
3	1	1	1	0	1
4	1	1	1	1	0
5	0	1	1	1	0
6	0	0	1	1	0
7	0	0	0	1	0
8	0	0	0	0	1
At the ninth pulse, conditions return to those of pulse 1.					

Apart from direct pulse counting, counter devices can perform many functions in industrial process control. Any transducer that can produce a pulse output can be used with a counter to automate an action on the basis of the number of counts received.

- **Downward count**: a downward count to zero is useful. This is used on some automatic coil-winding machines. The required number of turns to be wound on to a bobbin is first preset, and the machine set in motion. When the count reaches zero the machine automatically comes to a halt.
- **Frequency**: a counter can be used to measure the frequency of both unipolar (d.c.) and bipolar (a.c.) waveforms by counting the number of cycles over a precisely controlled period. It can also be used to measure time if driven by a precisely known crystal frequency.
- **Timekeeping**: this is the basis of the digital watch or clock. In radar or sonar systems a transmitted pulse is reflected from an object and some

of the energy returned to the source. The time delay can be measured and since the velocity of radio or sound waves is known, the time can be converted into distance.

*   **Velocity**: two photocell/light source pairs can be set up a known fixed distance apart. The time taken for an object to pass between them can be measured and the distance/time computed to give the velocity.

Earlier, we saw that switches produce multiple output pulses, called contact bounce, when switched. This is the reason for debouncing circuits.

Practical 37.4

Use a counter circuit and a switch without a debouncing circuit to show how many pulses that switch bounce causes. Try a variety of different switches.

Figure 37.14 shows a typical counting device designed to measure frequency. A crystal-controlled oscillator's output frequency is divided down to produce an accurate time gating signal. In practice, the frequency is chosen to be high enough to provide the maximum count or minimum time ranges to meet the user demand.

In the case shown, the crystal frequency is 32.768 kHz which, when divided by 2^{15}, provides a time gate of 1 s. The input signal is converted into a square wave by a suitable circuit and then used to clock the counters. The displayed count then represents the unknown frequency in cycles per second or hertz.

This circuit uses several different types of counter that have been covered in the preceding pages. The oscillator and first divider is a single IC 74HC4060. This contains a divide by 2^{14} ripple counter and provides a 2 Hz output to the 74HC4017, which is a mod-10 Johnson ring counter and is used to generate a sequence of control signals.

The COUNT signal is a 1 s wide positive pulse from the Q0 output of the 74HC4017, which enables the counters for exactly 1 s, in this case synchronous BCD counters. The Q2 output of the 74HC4017 is inverted to provide the active low LATCH signal for the 74HC4511 display drivers, which latches the count so that it can be displayed. The RESET signal is also inverted to drive the active low MR pin of the 74HC160 counter. The counters are reset ready for the next count, but the display continues to display the value that has been latched. The circuit updates once every 10 s, which is the time taken for the Johnson counter to go through a complete cycle.

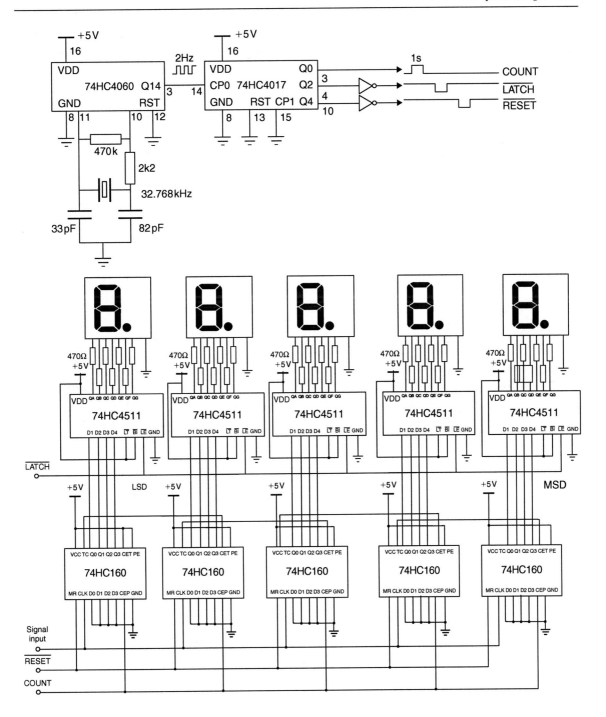

Figure 37.14 Simple five-digit frequency counter circuit

Multiple-choice revision questions

37.1 How is a transparent latch different from a D-type flip-flop?
 (a) it is edge triggered
 (b) it is level triggered
 (c) it does not have a preset input
 (d) it does not have a $\overline{Q}$ output.

37.2 How does the master–slave J-K flip-flop help in high-speed logic design?
 (a) it minimizes propagation delay between stages
 (b) J input connects directly to Q and K to $\overline{Q}$ of the previous stage
 (c) is faster because it is clocked internally on both edges of the clock signal
 (d) it avoids race conditions from the output to the input.

37.3 Why should ripple counters be avoided in decoded applications?
 (a) they are inefficient
 (b) race conditions exist as the count ripples through

 (c) they count in binary only
 (d) they use more stages to count than a Johnson counter.

37.4 What sort of shift register can be used for converting parallel data to serial stream?
 (a) SIPO
 (b) SISO
 (c) PISO
 (d) FIFO.

37.5 What is the main advantage of ring counters over binary ones when used as dividers?
 (a) they can divide by arbitrary integer values more quickly
 (b) they can be made from shift registers
 (c) they can count up as well as down
 (d) they have 50% duty cycle output.

38 Digital communications

The modern world is dependent on digital communications. Until relatively recently, radio, television and telephone systems were essentially analogue in nature, so, for example, speech carried by a telephone call was represented by continuously varying voltages or currents in proportion to the amplitude of the sound wave to be transmitted. Digital signals from devices such as computers and fax machines were in the minority. Today, however, most systems communicate digitally, including mobile phones, compact disc (CD) and MP3 players, and television systems such as FreeView. One important result of the move to digital communications is the convergence of devices such as televisions and video recorders with computers, or cameras and portable music players with mobile phones, into universal media devices.

The majority of communications, however, still originate in the physical world as sound and images, mainly voice (as in telephone communications), and such signals must be converted to some digital form before they enter the communications links. These communications may be by copper wire, by fibreoptic cable, or by modulated radio waves using either terrestrial or satellite broadcasting.

Practical 38.1

Connect two digital circuits by a fibreoptic link and verify that signals are being passed.

We class digital signals by their structure and information transfer rate. For convenience, binary digits (**bits**) are organized into various sized groups. For example, 4 bits form a **nibble** which can represent one hexadecimal character, and 8 bits form a **byte** or octet which can be used to represent an alphanumeric or control character from the ASCII (American Standard Code for Information Interchange) code set. A **word** may consist of several bytes, typically 2, 4 or 8 bytes in length. The data transfer rate may be quoted, depending on the application, in bits, bytes or characters per second.

The term **Baud rate** may be encountered, and this can cause confusion. The Baud rate for a communications systems means the **symbol rate**, where a symbol is the simplest modulating form used and, using suitable modulation techniques, it is possible that each transmitted symbol can represent more than 1 bit of information. If each symbol represents 4 bits, then the bit rate is four times the Baud or symbol rate. Only if each symbol represents 1 bit are the bit and Baud rates equal. Phase shift modulation methods can have a bit rate considerably higher than the Baud rate.

The term Baud rate is often used loosely and incorrectly (particularly in computing) to mean bits per second. For this reason, the term **symbol rate** is now more often used in digital television engineering applications.

Pulse code modulation

The form of digital signal that is usually used for speech data is pulse code modulation (PCM), meaning that each unit of data is represented by a binary number. Analogue signals are converted to PCM data by **sampling**, **quantizing** and **coding**.

Sampling means finding and storing the amplitude of a signal. By using sampling, the analogue signal is broken down into a set of time-related voltage levels (Figure 38.1), each of which can be stored until it is converted into a binary number that is proportional to the amplitude of the sample. Since the binary representation of the analogue number must make use of

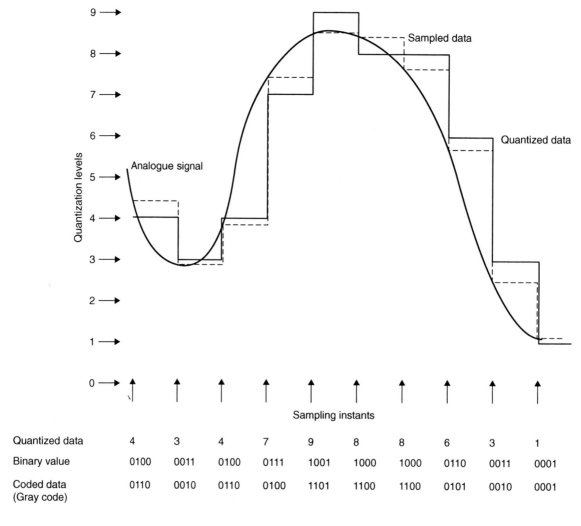

Quantized data	4	3	4	7	9	8	8	6	3	1
Binary value	0100	0011	0100	0111	1001	1000	1000	0110	0011	0001
Coded data (Gray code)	0110	0010	0110	0100	1101	1100	1100	0101	0010	0001

Figure 38.1   How PCM data may be derived from an analogue signal

a finite, often quite small, number of levels, the signal amplitude is said to be quantized. This quantized representation of the analogue signal is then coded to improve the robustness in communication and sometimes remove redundant data to make the digital communication as efficient as possible. The output of these sampling, quantizing and coding stages is the PCM data that can then be stored as a file of digital bits or transmitted along digital links. The use of Gray code, as shown in Figure 38.1, has advantages, in that in Gray-coded binary, successive numbers only differ by 1 bit; for example, the code for 5 is 0111, while 6 is represented by 0101. This means that the single-bit errors in communication have a relatively small effect on the output data compared with uncoded data.

Conversion of analogue signals into digital data and back again is key to the use of digital communications for services such as telephone and television. The notable features of this process are the sampling rate and the resolution; that is, the number of binary digits to represent the level.

The advantages of using digital communication include:

- It can provide a significantly higher transmission speed than can usually be achieved with analogue processing.
- It provides for improved transmission quality when electrical noise is present. Since the expected data can only take one of a number of pre-defined states, rather than being continuously variable as in analogue signals, the noise component can be reduced using error detection/correction techniques. This allows noiseless repeaters or signal regenerators.
- It is more compatible with the digital switching techniques used to control distribution and is a natural technique to use for systems involving an optical fibre link.
- Encryption/decryption can easily be added for data security, at any stage in the link because the characteristics of the signal are well known.
- Where necessary, signal compression/bit rate reduction techniques can be used to minimize the bandwidth requirement.
- For many applications, time division multiplexing (TDM) can be used more effectively than frequency division multiplexing (FDM), which is common for analogue transmission systems.
- For systems involving reception and retransmission, signal regeneration can be used at the intermediate stage to improve signal quality.

Time division multiplexing is a method for making efficient use of a communications channel (which may be a radio frequency band, a copper cable or a fibreoptic link) so as to carry more than one channel of information. Figure 38.2 shows a simplified example in which a pair of synchronized rotary switches selects the input and output data.

The principle is that a set of bytes from one channel is transmitted, followed by a set from another channel, and so on, until the process is repeated with bytes from the first channel again. This is possible only if:

- the bytes belonging to each channel can be identified at the receiver
- the bytes for any one channel can be assembled in memory at the receiver as fast as they are required

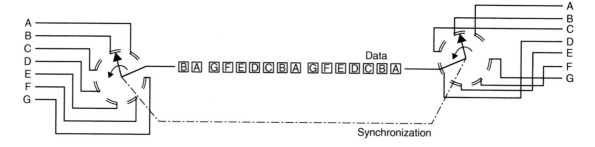

Figure 38.2   Simplified schematic of a time division multiplexed communication link

- the rate of sending bytes can be controlled so that the memory at the receiver does not overflow.

Schemes for TDM all depend on the use of **packets** of data. A **data packet** is a set of bytes of channel information, such as PCM from a sampled source, together with bytes that indicate the start and end of the packet, with identification to ensure that a packet intended for one channel cannot be confused with a packet for another channel.

In the telephone network the integrated services digital network (ISDN) provides a TDM based link between the customer premises and the local exchange. This is operated over the subscriber's local loop connection in place of an analogue telephone service. In contrast, broadband connections typically use asymmetric digital subscriber line (ADSL) and can be operated alongside the analogue telephone service.

The ISDN link can carry such services as telephony, telex, fax, private digital services and various speeds of circuit switched data service, up to 64 kbits/s. Figure 38.3 shows a typical packet structure for ISDN. The start flag indicates the beginning of the packet and can contain a sequence of bits designed to aid synchronization. This is followed by an address that identifies the package. For ISDN, this contains identification codes for the originator of the data and the destination of the data. The control section then contains the packet serial number to ensure that the packets are assembled into the correct order at the receiving end.

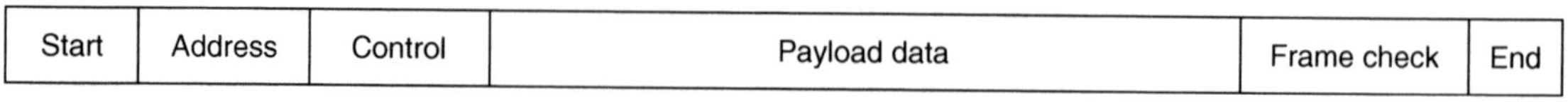

Start	Address	Control	Payload data	Frame check	End

Figure 38.3   ISDN packet structure

The payload data contains the message bytes, typically 256 bytes in each packet. The data bytes are followed by a frame check that is used to provide error protection and correction, and then an end flag to signify the last bit in the packet.

At the receiving end, logic circuits identify each packet and assign it to its correct channel. In each channel, the packets are assembled in memory and fed to their destination, which may be another storage device (such as a hard disk) or another converter such as a digital-to-analogue (D-A) converter that will convert the digital data stream into analogue format.

This system requires some form of control of transmission rate, so that the memory that is used for packet assembly (the **buffer**) cannot be overfilled. The usual scheme is that the rate at which the buffer assembles information is equal to the rate at which D-A conversion is carried out, and the number of channels being carried is calculated to supply the correct rate of packets in each channel. More complex methods are needed when the rate of reading packets may vary.

Practical 38.2

Use a TDM system to transmit several channels of data through a single path, and separate the channels again.

Digital data, encoded and multiplexed as packets, can be transmitted over fixed links, such as copper or fibreoptic cables, or by way of modulated radio waves. In general, some form of modulation of a carrier is used no matter how the data is transmitted, because modulation systems can be devised that allow very efficient use of the bandwidth of the transmission system (with a bit rate that is much higher than the Baud or symbol rate). The earlier systems used for digital modulation were **amplitude shift keying (ASK)** and **frequency shift keying (FSK)**.

Amplitude shift keying methods modulate the amplitude of the carrier according to the discrete level of the data signal. For binary modulation, the carrier is simply switched ON or OFF to represent 1 or 0, respectively. This technique is also known as **on–off keying (OOK)** (Figure 38.4). Because ASK and noise signals have similar characteristics (a change in signal level), this form of modulation is not used nowadays because of its poor signal-to-noise (S/N) ratio and hence a relatively high **bit error rate (BER)**.

Binary FSK involves switching the carrier wave between two set frequencies. If FSK is used with frequencies in the audio band, typically between 300 Hz and 3 kHz, it is described as audio frequency shift keying (AFSK). One form of this, much used for small computers that stored data on audiotape, is known as **Kansas City modulation**, for which bursts of eight cycles of 2400 Hz or four cycles of 1200 Hz are used to represent 1 or 0, respectively. Other frequency values can be used and these may or may not have an exact 2:1 ratio. Small differences are often introduced so that beat notes between the two frequencies will not introduce bit errors. Carrier frequency FSK has a better BER performance than ASK under the same conditions, but it requires a wider transmission bandwidth. The AFSK base band signals may be transmitted over cable or radio links by other techniques such as amplitude modulation (AM) or frequency modulation (FM).

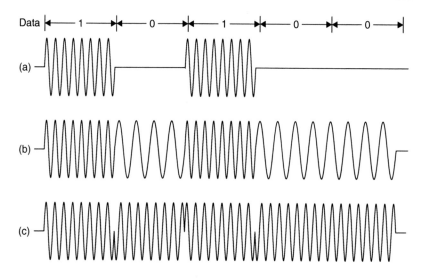

Figure 38.4   Digital modulation: (a) OOK, (b) FSK, and (c) PSK

Phase shift keying (PSK) is a more modern single-frequency modulation method in which the data signal is used to shift the carrier phase. Typically, for binary transmission a 0 produces no effect while a 1 generates 180° of carrier phase shift. Of the three methods we have looked at, PSK has the best BER performance and the narrowest transmission bandwidth, so modern methods of digital communications all make use of some form of PSK.

For cable transmissions, either copper or fibreoptic, **multiphase PSK**, in which the modulated carrier can carry a number of different phase shifts, can be used to advantage. Figure 38.5(a) shows the relationship between various amplitude and phase shifts and their vector diagrams. When both phase and amplitude modulation is used, the modulation is called **quadrature amplitude modulation (QAM)**. Figure 38.6(a) shows an example of 8-PSK where each vector can be used to represent 3 bits of information so that the Baud rate is just one-third of the bit rate. The concept of 4-PSK or quadrature PSK (QPSK) can be usefully extended in several ways. For example, if each of the four vectors is permitted to have any one of four different amplitudes, then each vector can be used to represent 4 bits. Figure 38.6(b) shows one example of such QAM (16-QAM), in which the signal points in the matrix form a constellation. Cable digital television systems can make use of even more bits of PSK, and one common system is 64-QAM.

Digital data streams containing long sequences of ones or zeros can be used within a computer, but they are not ideal for transmission. For example, a long stream of bits of the same value is rather like a very low-frequency analogue signal. This can cause problems with data transmission and recovery in receivers, and methods, classed as **secondary encoding**, have to be used to ensure acceptable performance. Similar problems are encountered with raw data recording on magnetic and other media.

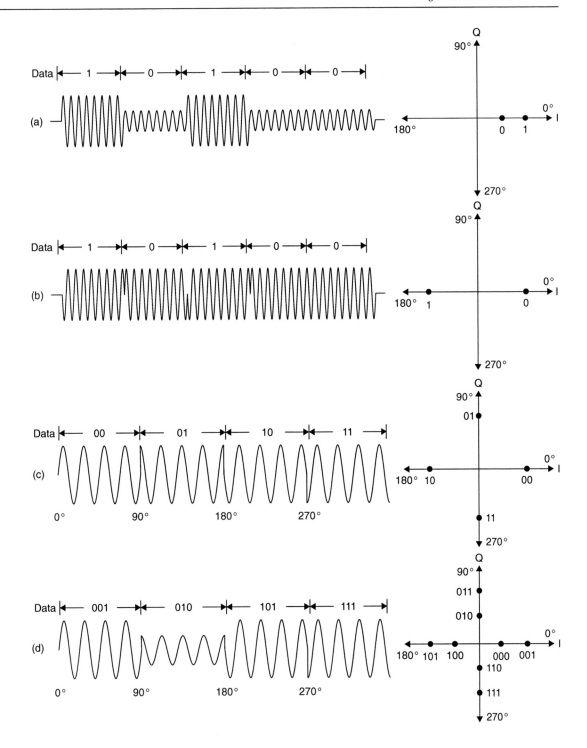

Figure 38.5   Multiamplitude and multiphase keying: (a) two-level ASK, (b) BPSK, (c) 4-PSK, and (d) 8-QAM

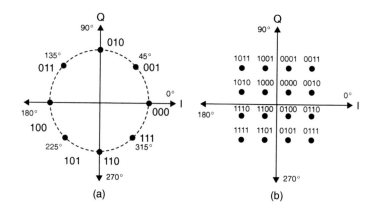

Figure 38.6   Digital modulation: (a) 8-PSK, and (b) 16-QAM

Digital communications systems are more robust than the equivalent analogue systems in the presence of noise. This is clearly seen by reference to Figure 38.7(a), which shows how even a relatively noisy signal can be recovered correctly. Figure 38.7(b) further compares analogue and digital signals under the same S/N ratio at the input.

While the analogue system degrades gracefully before it fails, the digital system is relatively unaffected by the noise until a point is reached, when the system suddenly crashes as the BER rises. However, the advantages of digital technology do not end here. There are a number of ways in which the information can be accurately recovered even from signals with a high BER, providing that the time scale is not critical.

At the simplest level of communication between two computers, a system known as automatic request for repeat (**ARQ**) is available. Reference to the ASCII code table shows that two special codes are available, **ACK** (acknowledge) and **NAK** (negative acknowledge). If a distant receiver detects a code pattern without errors, it transmits via a return channel the code ACK. However, if errors have been detected, transmission of the NAK code automatically generates a request for a repeat transmission of the last block of signal code.

Furthermore, bit errors can be detected and even corrected in a digital system using a technique known as forward error correction (**FEC**). This is so called because the means of error detection/correction are contained within the transmitted message stream. This is achieved by the addition of extra redundant bits which, when suitably processed, are capable of identifying the errors. Either method makes extra demands on the spectrum; FEC requires additional time or bandwidth to include the extra bits, while the ARQ system requires a free return channel.

The prime causes of bit errors are **white noise** and **impulsive noise**. White noise produces errors that are completely uncorrelated and random in occurrence, and impulsive noise creates a loss of bit stream synchronism, which leads to bursts of errors.

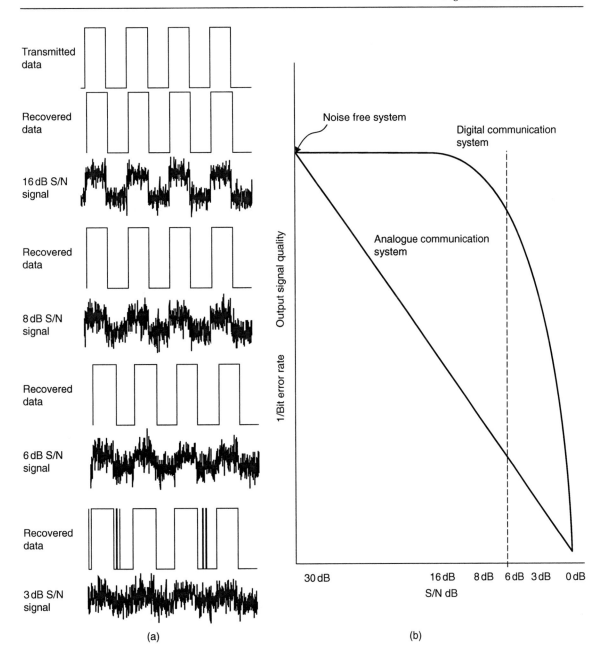

Figure 38.7   (a) Effect of noise on digital signal recovery, and (b) comparison of analogue and digital system behaviour in noisy conditions

Three classes of error need to be considered:

1. detectable and correctable
2. detectable but not correctable
3. undetectable and hence uncorrectable.

For any errors detected as type 2, the error can be concealed even if it cannot be corrected. The concealment options are:

- ignore the error and treat it as a zero level
- repeat the last known correct value
- interpolate between two known correct values.

Forward error correction can significantly enhance the resistance to noise interference for a digital communications system in a noisy environment. Comparatively simple methods can be applied to coded text that uses the **ASCII** code system. This 7 bit code allows for $2^7 = 128$ different alphabetic, numeric and control characters. The most commonly used digital word length is 8 bits or 1 byte ($2^8 = 256$), and therefore there is space for one extra redundant bit in each code pattern. Note that this spare bit is *not* available for extended ASCII codes (used by word-processing programs) that use all eight bits.

A single-error detection (**SED**) code of n binary digits is produced by placing $n - 1$ information or message bits in the first $n - 1$ bit positions of each word. The nth position is then filled with a 0 or 1 (the **parity bit**), chosen so that the entire code word contains an even number of 1s. For example, using an 8 bit byte to convey a 7 bit item of data allows the 8th bit to be used as a parity bit. If a code word that has been adjusted for **even parity** is received over a noisy link and is found to contain an odd number of 1s, then an error must have occurred.

Alternatively, a system may use **odd parity**, where the nth bit is such that the code word will contain an odd number of 1s. In either case, a parity check at the receiver will detect when an odd number of errors has occurred. The effect of suffering two (or any other even number) errors is self-cancelling, so these will pass undetected. The even or odd parity bits can be generated or tested using exclusive OR or exclusive NOR logic, respectively.

Such a pattern of bits is described as an **(n,k) code**, n bits long and containing k bits of information. The simple example of ASCII code is an (8,7) code. It thus follows that there are $n - k = c$ parity (or protection) bits in each code word. The set of 2^k possible code words is described as a block or linear code. ASCII uses $2^7 = 128$ words in a block code. At one time, all memory used in personal computers (PCs) allowed for 9 bit storage, so that a parity bit was available for checking memory integrity. This scheme is no longer used because memory is more reliable now, and there are better ways of checking memory integrity (such as by successive writing and reading methods).

The simple parity scheme that provides only a low level of error control is suitable only for systems that operate in relatively low levels of noise. The weak point of the system is that a single error may affect a parity bit, causing a valid byte to be rejected. In addition, the system is useful only in a communications link where a fault detected at the receiver can be signalled to the transmitter so that a block can be sent again.

The **check-sum** error control scheme is often used with magnetic storage media where data is stored in long addressable blocks. It is a simple scheme in which the digital sum of all the numbers in any block is stored at the end of each block. Recalculating the check sum and comparing the original and the new values after each data transfer quickly tests whether any read errors have occurred. For example, if the following binary words 0101 (5), 0011 (3), 1010 (10) and 0010 (2) are valid, then the check sum would be 10100 (20). If after a read operation the numbers became 0101 (5), 0011 (3), 1001 (9) and 0010 (2), the check sum should be 10011 (19). Comparison of the check sums shows that a read error has occurred and the block should be read again.

The simple check-sum technique can only identify when an error occurs and cannot indicate the position of the actual error. However, this problem can be overcome by using a **weighting** scheme that employs a series of **prime numbers** (any number that cannot be divided perfectly by any other number except for 1 and itself).

By multiplying each data number by a prime number, the check sum that is formed carries position information, and subtracting the true (transmitted) check sum from the false check sum (obtained by repeating the check-sum formation on the received faulty data) will give the multiplier number for the data digit that is in error.

This can be most easily explained using a series of denary numbers. Suppose we use the prime numbers 1, 3, 5 and 7, and the decimal values to be stored and read are 5, 3, 9 and 2. The weighted check sum would calculated by multiplying each data number by the corresponding prime number. In this example, it would be:

$$1 \times 5 + 3 \times 3 + 5 \times 9 + 7 \times 2 = 5 + 9 + 45 + 14 = 73$$

The sequence would thus be stored as 5, 3, 9, 2, 73. If on reading this became 5, 3, 8, 2, 73, the recalculated check-sum would be:

$$1 \times 5 + 3 \times 3 + 5 \times 8 + 7 \times 2 = 5 + 9 + 40 + 14 = 68$$

The check-sums are different, so an error has occurred, but the difference is $73 - 68 = 5$. Therefore a unit error has occurred in the 5-weighted value, the data number that was multiplied by 5 in this example, which was originally 9 and has become 8 because of the error. When this scheme is applied to a binary number, correction is simple, because if the digit 0 is an error, the true value is 1 and vice versa.

Much more advanced error correcting codes have been devised, and one type is named after **R. W. Hamming**, who was the originator of much of the early work on error control. Hamming code schemes are particularly useful for systems that require a high level of data integrity and operate in noisy environments that create random bit errors.

The cyclic redundancy check (CRC) system is most effective in combating burst errors. To generate the code for transmission, three code words are used. These are a **message code** word k, a **generator code** word G (selected to produce the desired characteristics of block-length and error detection/correction capability) and a **parity check code** word c. As for block codes, the transmission code word length is given by $n = k + c$. During encoding, the message k is loaded into a shift register and then moved c bits to the left, to make room for c parity bits. The register contents are then divided by the generator code word to produce a remainder that forms the parity check code word c, which is then loaded into the remaining shift register cells.

Thus, if the total code word were transmitted and received without error, this when divided by G, would yield a zero remainder. The last c bits can then be discarded to leave the original message code word. If however, an error occurs, then division by G leaves a remainder code word that acts as a syndrome. There is a one-to-one relationship between this and the error pattern, so that any correctable errors can be inverted by the error correcting logic within the decoder. The effectiveness of these codes depends largely on the generator code word, which has to be carefully selected. A further advantage is that CRC can be operated with microprocessor-based coding, when the system characteristics become reprogrammable.

Several other techniques have been developed from the basic Hamming concepts to deal with both random and bursts of errors. These include **BCH codes** for random error control, **Golay codes** for random and burst error control, and **Reed–Solomon codes** for random and very long burst errors with an economy of parity bits. Reed–Solomon coding is used for the CD recording system.

Primary and secondary codes

For a simple transmission system of number codes (**primary codes**) the BER can be minimized by using pulses of maximum width and/or amplitude, the obvious choice being a square shape. However, this introduces a number of problems. To pass a square wave, a transmission channel requires a wide bandwidth. To retain a good approximation to a square wave requires that the channel bandwidth should extend up to at least the 13th harmonic of the fundamental frequency. In any case, the transmission of such pulses through a typical channel will produce **dispersion** or pulse spreading, which leads to a type of error called **intersymbol interference** and an increase in the BER. Increased pulse width reduces the signalling speed and an increase in pulse amplitude introduces further problems. The final pulse shape is thus a compromise. One particularly useful pulse shape is described as a raised cosine. This shape is chosen because half of the pulse energy is contained within a bandwidth of half the bit rate.

A code format is a set of rules that defines clearly and with no room for mistakes the way in which binary digits can be used to represent alphabetic, numeric, graphic and control character symbols. The PCM code is one useful format, but other formats can be used with advantage to reduce the BER when noise is likely to corrupt a transmission. Such changes to the original set of codes are called **secondary codes**.

Shannon's rule states that the communication channel capacity, in bits/second, is related to the available bandwidth and the S/N ratio. It also shows that these two parameters can be balanced to maximize the channel capacity for an acceptable BER suitable for a particular service.

To take advantage of this tradeoff, binary code formats are redesigned by inserting extra bits into the data stream in a controlled way. Some of the formats used are shown in Figure 38.8. The general aim is to minimize the number of similar consecutive bits and balance the number of 1s and

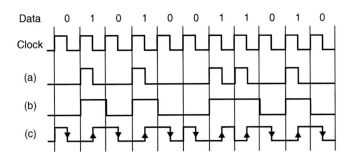

Figure 38.8 Commonly used code formats: (a) return to zero, (b) NRZ version, and (c) Manchester (biphase)

0s in the message stream. The greater number of signal transitions is used to improve the locking of the receiver clock and so reduce bit errors. The balance between the number of 1s and 0s produces a signal without a d.c. component in its power spectrum. This allows a.c. coupling to be used in the receiver and reduce its low-frequency response requirement. The commonly used codes are generated and decoded using dedicated integrated circuits [application-specific integrated circuits (ASICs)].

The return-to-zero (**RZ**) data, where a 1 is represented by a half-width pulse and a 0 by an absent pulse, is shown in Figure 38.8(a). This is a simple way of coding PCM, and it was used extensively in early telegraphy, but it is seldom used now because long runs of zeros can lead to synchronizing difficulties and because the reduced-width pulses require more energy to be transmitted per bit that in other coding methods.

Non-return-to-zero (**NRZ**) data is a signalling method in which a 1 is signified by a full-width pulse and a 0 by no pulse (Figure 38.8b). Long runs of 1s can lead to d.c. offset in the data communication system, which can cause problems in data transmission and recovery; long runs of 0s can also lead to synchronizing difficulties.

The **Manchester code** shown in Figure 38.8(c) is one of a number of biphase data formats, which are popular because they contain both clock and data information, making them useful for remote control applications, in which each bit in the original signal is represented by the direction of a transition in the centre of the bit period. The Manchester code is most easily derived by XORing the data and clock signals, as can be seen by studying Figure 38.8. Another way of looking at Manchester data is to observe that each bit is represented by a 2 bit long symbol; the transform is that 0 is represented by 10 and 1 by 01. Biphase data ensures that there are never more than two identical bits in series; this avoids the problems of d.c. offsets and synchronization suffered by both RZ and NRZ codes. Another variant of the Manchester format uses the opposite pair of symbols.

Another modulation format it that used by CD systems **EFM** (eleven-to-fourteen modulation) as a way of overcoming the problems of recording on CD media. In this system, the 8 bit code has been extended to 11 bits with parity bits, and each 11 bit number is transformed into a 14 bit number for recording. The 14 bit numbers are chosen so as to avoid the repetition of a digit or the rapid changes between 0 and 1 that can be troublesome in a recording or replay action.

The pure binary, 8-4-2-1, type of code is not the only binary code that we can use, and several others are likely to be encountered, depending on the type of applications. One very common variant of the 8-4-2-1 code is binary-coded decimal (**BCD**). This is a very popular form of coding when numbers have to be shown on liquid crystal display (LCD) or light-emitting diode (LED) displays, because each digit of such a display is one denary digit. It is also frequently used in process control equipment for the same reason, and some microcontrollers have BCD arithmetic instructions, which greatly simplify handling BCD numbers.

In a BCD system, then, each denary digit is represented by 4 bit binary code. Table 38.1 demonstrates what this implies. The conversion between BCD and denary is simple, and conversion to a form suitable for driving a display is also simple. The conversion between BCD and binary is, however, not quite so simple for numbers of more than one digit, and arithmetic with BCD is also not so simple. An added disadvantage is that BCD requires more storage space for any given number than 8-4-2-1 binary, although it is possible to devise systems in which number accuracy of floating point numbers can be much better, at the cost of very much slower arithmetical processes.

Table 38.1   Denary, hexadecimal, binary, BCD and Gray code equivalents

Denary	Hex	Binary	BCD	Gray
0	0	0000	0000	0000
1	1	0001	0001	0001
2	2	0010	0010	0011
3	3	0011	0011	0010
4	4	0100	0100	0110
5	5	0101	0101	0111
6	6	0110	0110	0101
7	7	0111	0111	0100
8	8	1000	1000	1100
9	9	1001	1001	1101
10	A	1010		1111
11	B	1011		1110
12	C	1100		1010
13	D	1101		1011
14	E	1110		1001
15	F	1111		1000

BCD is really just an adaptation of 8-4-2-1 binary, but Gray code is quite a different form. As Table 38.1 shows, a Gray code scale does not use columns to indicate the value of a bit, and you always have to use a table to convert a number in Gray code to a denary or a binary 8-4-2-1 number. The Gray code numbers are for 4 bits only, because Gray code is used either in

BCD form, or in a scale of 16 (hexadecimal). The Gray code is sometimes called the binary reflected code.

The advantage of Gray code is that only one bit ever changes at a time during a count-up. The change from 7 to 8, for example, is from 0100 to 1100, rather than from 0111 to 1000 in the 8-4-2-1 binary code. Gray code has particular advantages for conversion of quantities such as the rotation of a shaft into binary form, because if the shaft is in a position between the angles represented by numbers 7 and 8, then there may be reading errors caused by some binary digits that are changing between 0 and 1, and the fewer digits that change, the lower the chances of error.

There are other forms of binary code, such as **Excess-3**, which is a form of BCD in which 3 is added to each digit before coding into 8-4-2-1 binary. This means that the smallest value of code is 0011 and the largest is 1100, with anything below 0011 or above 1100 being an error. This makes error detection easier, and also has the considerable advantage that BCD numbers coded in this way can be manipulated by the same circuits as ordinary binary. The Gray code and the various forms of 8-4-2-1 code are, however, by far the predominant methods of coding that you are likely to encounter.

Multiple-choice revision questions

38.1 A data link transmits 8 bit data and has a Baud rate of 300 symbols per second. If the data is encoded as 1 bit per symbol with a start bit a parity bit and 2 stop bits per frame, what is the data bit rate of the link?
   (a) 300 bits/s
   (b) 200 bits/s
   (c) 120 bits/s
   (d) 100 bits/s.

38.2 A 16-QAM data link operates at 100k symbols per second. What is the bit rate of the raw data stream. (i.e. including error correcting codes and packet control bits)?
   (a) 100 kbits/s
   (b) 800 kbits/s
   (c) 1.6 Mbits/s
   (d) 6.4 Mbits/s.

38.3 What is the effect of using even parity on the bits in an *n* bit message block?
   (a) there will be an even number of zeros
   (b) there will be the same number of zeros as ones
   (c) there will be an even number of ones
   (d) there will be more ones than zeros.

38.4 What are the main advantages of using Manchester code to transmit data?
   (a) it increases the data rate and symbol rate
   (b) it minimizes d.c. signal offsets and maintains synchronization
   (c) it reduces the energy required and minimized d.c. offsets
   (d) it only transmits when bit values change and increases the data rate.

38.5 How would the decimal number 127 be represented in BCD?
   (a) 0001 0010 0111
   (b) 0111 0001 0010
   (c) 1000 1001 1110
   (d) 0000 0111 1111.

38.6 The main feature of TDM systems is that:
   (a) data is transmitted as soon as it is available to the system
   (b) data is transmitted in parallel packets using multiple pairs of wires
   (c) multiple data channels are transmitted sequentially over one physical channel
   (d) data is transmitted over fibreoptic links using modulated laser diodes.

39 Test equipment and fault-finding

General fault-finding methods

Fault-finding for electronic/electrical equipment is a skill that is neither an art nor a science, but an engineering discipline in its own right. Effective fault-finding requires:

- a good general knowledge of electricity and electronics.
- specialized knowledge of the faulty equipment.
- suitable test equipment.
- experience in using such test equipment.
- the ability to formulate a procedure for isolating a fault.
- the availability of service sheets and other guides.

A good general knowledge of electricity/electronics is essential because not all equipment is well documented, and in some cases only a circuit diagram (or even nothing at all) may be available as a guide. Failing a concise description of how the equipment works, you may have to work out for yourself the progress of a signal through the equipment. In addition, a wide general knowledge is needed if you are to make reasonable assumptions about how to substitute components. You are not likely to know why something does not work if you do not know what *does* make it work.

Specialized knowledge can greatly reduce the time spent in servicing, and if your servicing is confined to a few models of equipment you are likely to know common or recurring faults by their symptoms. All too often, however, service engineers are likely to have to struggle with unfamiliar equipment for a large portion of their time, and the services that are now available on the Internet allow you to plug into the experience of others.

Suitable test equipment is essential. The days when a service engineer could function effectively with little more than an multimeter and a screwdriver are long gone, and although the multimeter is still an important tool (as also is the screwdriver) the service engineer needs at least one good general-purpose oscilloscope, along with signal generators, pulse generators and more specialized equipment appropriate for the type of equipment being worked on.

Experience in using test equipment is also essential. All test instruments have limitations, and you must know what these are and how you can avoid being hung up by these limitations. You must know which tests are appropriate for the faulty equipment, and what the result of such tests would be on equipment that was not faulty.

The ability to formulate a procedure for isolating a fault means that you need to know what to test. All electrical and electronic equipment consists of sections, and much modern electronics equipment uses a single integrated circuit (IC) per section. You should be able to pin down a fault to one section in a logical way, so that you do not waste time in performing tests on parts of the circuit which could not possibly cause the fault.

An important point about all fault-finding is that you should make note of all the tests you have applied and the results of these tests. Without this, you are likely to forget what you have done, and end up taking measurements haphazardly, making diagnosis impossible.

Do not forget the simple things. We tend to make fun of the customer who has forgotten to check whether the mains switch was on, but we can be guilty of similar neglect ourselves. In particular, always check plugs and connectors for any equipment that is not functioning. The next logical check is that power supplies are present within the equipment. Once these elementary checks have been made, we can start looking for more serious faults.

The classical method of isolating a fault has, in the past, been to check signal inputs and outputs for each stage, but this is no longer the only method that needs to be used, and in some cases, the use of feedback loops, limiters and other interacting circuits makes it much more difficult to find where a fault lies. Once again, experience is a valuable guide.

The availability of service sheets and other guides is also important. Much commercial equipment consists of components that carry only factory codes, and whose actions you can only guess at in the absence of detailed information. In addition, good service sheets will often carry a list of known recurring faults, and will also give valuable hints on fault-finding methods.

Basic fault-finding in both analogue and digital systems follows principles that are similar. A source is required to inject suitable signals into the input and the signal processing is then monitored as it passes through the system on a stage-by-stage basis. For analogue systems a suitable input source is a signal generator, while an oscilloscope can be used as a monitor. For digital systems this **end-to-end** technique can be carried out using a logic pulser to provide the inputs while the processing can be monitored with a logic probe. You can use this method in either direction, input-to-output or output-to-input.

Another technique that is often used with advantage to speed up diagnosis is known as the **half-split** method (or **divide and conquer**). Here, the system is divided into two sections and the end-to-end technique used to find the faulty half. Once this is found, this part is again divided into two, and the check repeated. This process is then repeated continually until the faulty stage is identified. It is unusual to require more than three repetitions of this method.

Finding a fault is not, unfortunately, a certain step towards repair. Some equipment carries ICs which are no longer in production and for which no replacement is available. Many firms, particularly manufacturers of domestic electronic equipment, will provide spares and help for only a limited period, and some firms seem to deny all responsibility for what their equipment does after a few years. Given the comparatively long life of most electronic equipment, it would be unreasonable to expect spares to be available indefinitely, but it is not easy to tell a customer that the television receiver bought only 6 years earlier cannot now be serviced because it contains parts for which these is no current equivalent. Manufacturers may like to remember that customers tend to have long memories about such things: it certainly affects my judgement when I want to buy anything new.

Servicing digital systems

In some respects, servicing logic circuitry can be simpler than working on analogue circuits of comparable size. All digital signals are voltages at one of two levels, and there is no problem of identifying minor changes of waveshape which can so often cause trouble in linear circuits. In addition, the specifications that have to be met by a microprocessor circuit can be expressed in less ambiguous terms than those that have to be used for analogue circuits. You do not, for example, have to worry about harmonic distortion or intermodulation, and parasitic oscillation is rare, although certainly not impossible.

That said, logic and microprocessor circuits bring their own particular headaches, the worst of which are the relative timing of voltage changes and the difficulty of displaying signals. The actions of any microprocessor circuit depend on strict timing of many pulses being maintained, and conventional equipment which serves well for the analysis of analogue circuits is of little use in working with microprocessor circuits. The problem is compounded by the fast clock rates that have to be used for many types of microprocessor.

You may, for example, be looking for a coincidence of two pulses with a 66 MHz clock pulse, with the problem that the coinciding pulses happen only when a particular action is taking place. This action may be completely masked by many others on the same lines, and it is in this respect that the conventional oscilloscope is least useful. Oscilloscopes as used in analogue circuits are intended to display repetitive waveforms, and are not particularly useful for displaying a waveform which once in 300 cycles shows a different pattern. A fast conventional oscilloscope is useful for checking clock pulse rise and fall times, and for a few other measurements, but for anything that involves bus actions a good storage oscilloscope is needed.

Software and hardware, such as that provided by Pico Technology, can be used along with a fast modern personal computer (PC) to provide facilities for logic circuit diagnosis (see Chapter 31).

In addition, some more specialized equipment will be necessary if anything other than fairly simple work is to be contemplated. Most of this work is likely to be on machine control circuits and the larger types of computer. Small computers do not offer sufficient profit margin in repair work to justify much diagnostic equipment. After all, there is not much point in carrying out a £300 repair on a machine that is being discounted in the shops to £399! We shall start this chapter looking at some typical fault conditions, and then at specialized instruments for digital circuit work.

Practical 39.1

Use suitable test equipment to locate faults in remote controls, A-D and D-A conversion circuits, and seven-segment displays.

Common problems

Some of the problems that occur in digital equipment are due to poor servicing or poor manufacturing techniques, while others may be brought about by failure of associated equipment (such as a regulated power supply). In this section we shall concentrate on common causes of failure of ICs. These are:

- incorrect insertion into sockets
- pins shorted during measurements
- poor soldering techniques
- insertion/removal with power applied
- incorrect voltage levels
- input pins left disconnected
- electrostatic discharge.

Incorrect integrated circuit insertion

Incorrect insertion into a socket is most likely for the larger dual in line (DIL) ICs, and the most common error is that pin 1 of the IC has been put into the pin 3 position of the socket, or pin 3 of the IC has been put into the pin 1 position of the socket. Since the V+ and earth pins are usually located at opposite corners of a chip, incorrect insertion will almost certainly make one of these connections open-circuit. The older Schottky transistor–transistor logic (STTL) chips (and many of the LS types) will survive having power switched on in these circumstances, but metal oxide semiconductor (MOS) chips are more vulnerable. Always be prepared to test and, if necessary, replace a chip that has been incorrectly inserted.

Live measurements

Making measurements of voltages on a (live) working chip is reasonably simple for chips of the 74 family type, provided the correct tools are used. By far the best method is to slip an adapter over the chip. These are made for the most popular pin layouts, and they provide a set of contacts on the top cover that correspond to the IC pins and are electrically connected through clips. These contacts are well away from the printed circuit board (PCB) surface, and you can make clip connections to them without any risk of short-circuits.

Without such adapters, great care has to be taken over pin voltage measurements, particularly the voltage difference between pins. Absolute voltage readings can be taken by clipping one lead of the meter to an earth point, and using a probe held to the IC pin. The probe should be insulated almost to the contact point to prevent shorting between pins.

By far the worst measuring problems relate to closely spaced pins, particularly the square layouts used on microprocessors and application-specific integrated circuits (ASICs). Adapters are available from the manufacturers for some types, but the only alternative is to use a fine probe with insulation almost to the tip.

Poor soldering

Poor soldering is a problem that is not quite so common now that manufacturing techniques have improved, but it is still a source of faults. The nature of modern boards with very fine tracks, small solder-pads and surface-mounted (SM) components requires excellent soldering systems, and poses a problem for fault-finding because a badly soldered joint is very difficult to detect by eye.

Another, more recent, problem is that the use of modern lead-free solders seems to have caused a rise in the incidence of dry joints.

A good magnifying glass and a bright light can help, but if you are convinced that the problem is in an area that appears to be all right, the usual treatment is to apply a soldering iron to see whether this clears up the problem. The iron should be fitted with a very small bit, and it often helps to give the suspect area a thin coating of an approved flux paint. Needless to say, the equipment must be switched off and time allowed for capacitors to discharge before any such resoldering is attempted.

Another aspect of this problem is a poorly soldered repair. Even with care, some repairs on modern PCBs are very difficult, particularly with SM components, and soldering must be suspected if a circuit fails to operate correctly after a repair (involving soldering) has been carried out.

Power off

We tend to take it for granted that power will be off when any repair is being done, but there is a temptation to remove and insert chips with power still on. This is something that must be resisted, even if switching off and on is not a straightforward action (as for many computer boards). Not only must the power be off when any chip is removed or replaced, it must have been off for long enough for capacitors to have discharged.

Remember that some PCBs may contain backup batteries, and these may have to be temporarily removed for some repairs. Since the backup batteries are there to keep memory working, this will require a considerable amount of work after restoring the battery, and you should consult the manufacturer's recommendation on this subject. If you need to replace a battery on a PC motherboard, you may be able to connect an external battery to maintain the memory while the internal cell (nowadays often a lithium-ion type) is replaced. If not, motherboards usually provide for a set of defaults to come into use after the backup battery power has been interrupted.

Incorrect voltage levels

Incorrect voltage levels can arise in several ways. One is the incorrect connection of cables that carry power from a power supply unit (PSU) to a board, and another is the failure of regulation of a power supply. Incorrect cable connection is unlikely on units such as modern PCs (using the ATX motherboard system), but if there is any doubt, the alignment of the connectors should be marked before they are separated.

Failure of regulation is a problem that is less easily avoided. Some circuit boards are surprisingly immune to regulation failure, particularly when no signals are applied, but for complex circuits, the PSU should incorporate overvoltage protection systems.

Any work that is done on the PSU should preferably be done with the unit separated, and then with a dummy load connected for testing. In this way, any faults that cause incorrect voltage outputs will not damage ICs on the main boards.

Disconnected inputs

Input pin disconnections can arise from the incorrect insertion of ICs into sockets or from incorrect soldering of ICs to a board. They cause little harm on STTL circuits, where a disconnected input will float high, but can cause damage on MOS ICs when power is applied. Modern circuit boards are designed so that a path to earth is available on any input, but only if the IC pins are connected to their tracks.

Electrostatic discharge

Electrostatic discharge (ESD) should never be a problem for any IC that is correctly connected into circuit on a board, because the board will provide discharge paths. Damage is more likely when ICs are handled before insertion in a board, because the only protection available is by way of diodes incorporated into the chips, and these offer only limited protection.

The comparatively high humidity of the climate in the UK allows us to get away with working practices that would cause electrostatic damage elsewhere, and it is quite common even in manufacturing to see MOS devices being handled without elaborate electrostatic precautions. This is not, however, good practice, and the basic precautions of working in a well-earthed environment and keeping one hand earthed should be observed.

The worst hazards are the ever-present plastic bags and other plastic packaging materials, all of which are electrostatic hazards. Computer components are always packed in conductive plastic (brown or black), and a collection of these materials can be useful to provide a clean conducting surface for placing sensitive ICs.

Electrostatic damage, see also (Chapters 19 [Level-2 Book] and 28) is most likely to affect metal oxide semiconductor field-effect transistors (MOSFETs), some of which can be damaged by a voltage as low as 35 V. Damage to a simple transistor is likely to cause failure, but damage to elaborate MOS ICs may be less obvious, causing reduced performance and/or shortened life. Table 39.1 indicates the level of electrostatic voltages that

Table 39.1 Electrostatic voltages in the workshop

Action	*Relative humidity*	
	Low (10–20%)	*High (65–90%)*
Walking over carpet (artificial fibres)	35 kV	1.5 kV
Walking on vinyl floor	12 kV	250 V
Working on unprotected bench	6 kV	100 V
Shuffling paper	7 kV	600 V
Picking up polythene bag	20 kV	1.2 kV

can be generated under typical conditions. Note that even bipolar devices and diodes can be damaged by less than 3 kV levels, although the amount of current that can pass is often too low to cause lasting damage.

The figures for high relative humidity apply fairly generally in the UK, where the workshop is naturally ventilated, but the lower figures are more likely where air-conditioning is used.

Damage to MOS ICs is possible whenever an electrostatic voltage can be applied to a pin or set of pins and can cause current to flow to grounded pins. This is most likely if an IC is incorrectly inserted or soldered so that some pins are earthed and one or more is unconnected.

Device sensitivity

The devices that are most sensitive to electrostatic damage are MOS and complementary metal-oxide semiconductor (CMOS) chips that have no internal protection circuits, and discrete MOSFETs that also have no discharge paths built in. These devices are very rare now, but if they are present the utmost caution should be exercised in handling the devices, observing all the precautions recommended by the manufacturers.

Less sensitive devices include 74LS chips, analogue ICs, r.f. transistors for 500 MHz or more, semiconductors that incorporate silicon dioxide insulation, and MOS/CMOS devices with diode protection for inputs, along with junction field-effect transistors (JFETs) and precision thin-film resistors. The least sensitive devices include microcircuits, small-signal transistors (less than 10 W output) and thick film resistors.

To put all this into perspective, I have never had a chip damaged by ESD in 45 years of handling components. Damage to items such as low noise block (LNB) circuits is more likely to be caused by electrical storms.

Some of the protective measures that are normally used to prevent ESD damage are noted here.

• Before starting any servicing work, check technical manuals and any leaflets for warnings and instructions regarding electrostatic damage precautions. Ensure that the working environment is as free as possible from electrostatic hazards, and try to ensure that nothing you do will generate high electrostatic voltages (e.g. shuffling your feet on a carpet or on vinyl tiles).

• Before servicing equipment, you should earth yourself. The simplest way is to touch an earthed metal object, but if you have acquired a substantial amount of charge this can be painful. A better method is to provide in the workshop an earthed rail or wire to which 1 M resistors are connected at intervals, with a small metal ball soldered to the free end of each resistor. Touching one of these metal balls will earth you painlessly without sparks.

• In low-humidity conditions it is much more satisfactory to wear an earthing strap, a metal strap that is permanently connected (usually through a 1M resistor) to earth.

Safety regulations demand that you should not use any mains-operated equipment while you are wearing an earthing strap.

The most risky procedures involve the unpacking of new devices. Handling must be kept to a minimum, with attention to earthing at all times, and you should try to ensure that the workshop environment is maintained so that materials that cause high electrostatic voltages are kept to a minimum. Do not unpack sensitive devices until you are ready to use them.

The conductive packing plastic that is used for sensitive devices can be used to hold the chips while you insert them, but if this is not possible, hold ICs by the body corners only, avoiding any contact with the pins. Avoid in particular any contact between IC pins and any clothing or (non-conducting) plastic. If you are not wearing a wrist strap, try to keep contact with an earthed surface while you are handling a sensitive IC. Avoid touching synthetic materials that acquire electrostatic charges. Use soldering irons that are earthed, and avoid the use of solder suckers that have plastic (PTFE) tips, unless the tips are guaranteed to be antistatic.

Practical 39.2

Use suitable test equipment to locate faults in synchronous and asynchronous counters, shift registers and bistable circuits.

Test equipment

Test equipment for digital circuitry must, of necessity, include some items that are not normally used on analogue circuits, but this does not mean that familiar instruments such as the multimeter are used to any less an extent. A good analogue or digital voltmeter is a very valuable backbone of all modern servicing, but you should know its limitations.

When the analogue meter is used for voltage readings you need to know the resistance of the meter for each voltage range. The meter resistance can be found from the ohm-per-volt figure, or figure-of-merit, which is printed either on the meter itself or in its instruction booklet. To find the resistance of the meter on a given voltage range, multiply the figure-of-merit by the voltage of the required range.

Example: What is the resistance of a 20 KΩ/V voltmeter on its 10 V range?

Solution: Meter resistance is 20 K × 10 = 200 K on the 10 V range.

If you need to take a reading across a high resistance, a high-resistance meter range must be used. You may be able to use a higher meter range than the one that seems to be called for. For example, if a voltage of around

9 V is to be measured and the 10 V range of the meter has too low a resistance, the 50 V or even the 100 V range can be used to reduce the distorting effect of the meter on the circuit. Note, however, that this would not be possible if the reading on the 100 V range would thereby be made too low to read. A voltage of around 1.5 V or less would be unreadable on the higher range scales.

Digital multimeters generally have much less effect on the circuit being measured. Most of them have a constant input resistance of about 10 M or more, and few circuits will be greatly affected by having such a meter connected into them. The operating principle is that the input voltage is applied to a high-resistance potential divider which feeds a comparator.

- In general, digital meters are not capable of following rapid changes in voltage.

- Although a digital meter may indicate a voltage reading to several places of decimals, this is not necessarily more precise than the reading on an analogue meter.

- Meters with still higher resistances are also available up to several thousand megohms, if required.

Current tracer

Work on PCBs makes it difficult to trace currents, since it would be unacceptable to split a track to check the current flowing through it. A current tracer (also called a current checker) is a small handheld device that makes it possible to detect the presence and direction of current flow on a PCB track, and also to indicate, roughly, its size.

The principles of one type are illustrated in the block diagram of Figure 39.1. Two probes, set a fixed distance apart, are pressed on the track, and the tiny voltage difference that is produced by the current flow is amplified by balanced operational amplifiers so as to operate light-emitting diode (LED) indicators. Typically, three LEDs are used to indicate currents of 10, 50 and 100 mA on 1 mm track width, or 20, 100 and 200 mA on 2 mm track width. A fourth LED is used to indicate reverse polarity, or to indicate open-circuit track.

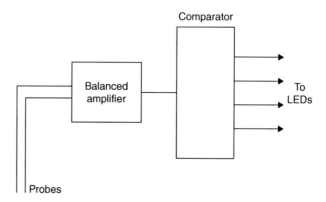

Figure 39.1   Block diagram of one type of current tracer

An alternative method of tracing current uses a Hall-effect probe which does not rely on sensing voltage, using instead the magnetic field around the track. This type of tracer, such as the ToneOhm, manufactured by Polar Instruments, will detect partial and complete short-circuits and, with practice, can be used to find the position of an open-circuit.

A current tracer, like a voltmeter, is a useful first test instrument, as it will assist in finding any problems that cause a drastic change in the current flowing along a track. This can help to pinpoint damaged tracks (a hazard particularly in flexible circuit strips) or failure of components that draw large currents, such as printer-head driver transistors. One point to note, however, is that modern PCBs with narrow tracks make the use of any type of current tracer very much more difficult.

Logic probe

A logic probe is a device which uses a small conducting probe to investigate the logic state of a single line. The state of the line is indicated by LEDs, which will indicate high, low or pulsing signals on the line. The probe is of very high impedance, so the loading on the line is negligible.

A typical logic probe can be switched to either TTL or CMOS voltage levels ($+5$ V supply for TTL and 3–18 V for CMOS) and uses coloured LEDs to identify the logic state of the track. Pulses as narrow as 30 ns and as wide as 500 ms can be detected. Some types can also indicate the presence of ripple on the power supply, indicating faulty stabilization. Probes can be battery powered, or can take their power from the circuit under test, using crocodile clips.

These probes are not costly, around £15–50, and are extremely useful for a wide range of work on faults of the simpler type. They will not detect problems of mistiming, but such faults are rare if a circuit has been correctly designed in the first place. Most straightforward circuit problems, which are mainly chip faults or open- or short-circuits, can be discovered by the intelligent use of a logic probe, and since the probe is a pocket-sized instrument it is particularly useful for on-site servicing.

The probe, like the voltmeter used in an analogue circuit, has to be used along with some knowledge of the circuit. You cannot expect to gain much from simply probing each line of an unknown circuit. For a circuit about which little is known, however, some probing on the pins of the microprocessor can be very revealing. Since there is a limited number of widely used microprocessor types, it is possible to carry around a set of pinouts for all the microprocessors that will be encountered.

Starting with the most obvious point, the probe will reveal whether a clock pulse is present or not. Quite a surprising number of defective systems go down with this simple fault: it is even more common if the clock circuits are external. Other very obvious points to look for are a permanent activating voltage on a HALT line, or a permanent interrupt voltage, caused by short-circuits. For an intermittently functioning or partly functioning circuit, failure to find pulsing voltages on the higher address lines or on data lines may point to microprocessor or circuit-board faults.

For computers, the description of the fault condition along with knowledge of the service history may be enough to lead to a test of the line that is at fault. The considerable advantage of using logic probes is that they do not interfere with the circuit, are very unlikely to cause problems by their use, and are simple to use. Some 90% of microprocessor system faults are detectable by the use of logic probes, and they should always be the first hardware diagnostic tool that is brought into action against a troublesome circuit.

Logic clips and pulsers

The logic clip is an extension of the logic probe to cover more than one line. As the name suggests, these devices clip over a logic IC, and are available for 14- or 16-pin DIL packages. Each pin of the IC is connected to a buffer in the logic clip, and this buffer drives an LED indicator. Logic clips are usually available in separate TTL or CMOS versions, although the TTL version is now more common because of the widespread use of 74HC devices.

The logic clip is particularly useful when the system clock rate can be slowed down, or a logic pulser is being used to supply an input. Because all the states on a single IC chip can be monitored together, any fault in a gate within the chip is fairly easy to find, much easier than the use of a logic probe on all signal pins in turn.

Logic pulsers (or digital pulsers) are the companion device to the logic probe. Since the whole of a microprocessor system is software operated, some lines may never be active unless a suitable section of program happens to be running. In machine control circuits in particular, this piece of program may not run during any test, and some way will have to be found to test the lines for correct action.

A logic pulser, as the name indicates, will pulse a line briefly, almost irrespective of the loading effect of the chips attached to the line. The injected pulse can be detected by the logic probe. This method is particularly useful in tracing the path of a pulse through several gate and flip-flop stages. A synchronizing pulse can be used to trigger an oscilloscope if needed.

The logic pulser is a more specialized device than the logic probe, and it has to be used with more care. It can, however, be very useful, particularly where a diagnostic program is not available, or for testing actions that cannot readily be simulated.

Logic analyser

The logic analyser is an instrument which is designed for much more detailed and searching tests on digital circuits in general and on microprocessor circuits in particular. As we have noted, the conventional oscilloscope is of limited use in microprocessor circuits because of the constantly changing signals on the buses as the microprocessor steps through its program. Storage oscilloscopes allow relative timing of transitions to be examined for a limited number of channels, but suitable triggering is seldom available. Logic probes and monitors are useful for checking logic conditions, but are not helpful if the fault is one that concerns the timing of signals on different lines.

The logic analyser is intended to overcome these problems by allowing a time sample of voltages on many lines to be obtained, stored, and then examined at leisure. Most logic analysers permit two types of display. One is the timing diagram display, also called timing domain analysis, in which the various logic levels for each line are displayed in sequence, running from left to right on the output screen of the analyser. A more graphical form of this display can be obtained by connecting a conventional oscilloscope, in which case, the pattern will resemble that which would be obtained from a 16-channel storage oscilloscope. The synchronization may be from the clock of the microprocessor system, or at independent (and higher) clock rates which are more suited to displaying how signal levels change with time.

The other form of display is word display or data domain analysis. This uses a reading of all the sampled signals at each clock edge, and displays the results as a 'word' for each clock pulse rather than as a waveform. If the display is in binary, then the word will show directly the 0 and 1 levels on the various lines. For many purposes, display of the status word in other forms, such as hex, octal, denary or ASCII, may be appropriate. This display, which gives rise to a list of words as the system operates, is often better suited for work on a system that uses buses, such as any microprocessor system.

The triggering of either type of display may be at a single voltage transition, like the triggering of an oscilloscope, or it may be gated by some preset group of signals, such as an address (a trigger word or event). This allows for detecting problems that arise when one particular address is used or one particular instruction is executed. One common method is to trigger a display by using a combination of inputs. One of these would be a trigger word which can be set and stored, the other inputs would be trigger signals (qualifiers) which can be taken from the clock (a clock qualifier) or from other inputs (trigger qualifier). You can also use a word search (trace word or event action) through the memory of the analyser to find whether a specified word has been stored in the course of an analysis.

The data acquisition portion of an analyser samples the logic levels at the 16 inputs, using the internal or an external clock pulse to synchronize the sampling. These levels are stored, and in the usual operating mode, the triggering will cause the stored data to be centred around the triggering time. For example, there may be 2K of data both before and after the trigger event. Triggering is an ANDed action, so you can set for some combination of signals that is unique, such as when the microprocessor writes a specified word to a chip during an interrupt.

Note that a display such as can be achieved using a logic analyser can also be obtained from a computer simulator. Simulator software allows the user to notify the chips and connections that are used in a digital circuit, and the program will then provide simulated waveforms. The value of this system is that any unwanted pulses (called glitches) can be detected, even to the extent of a 1 ns pulse that in real life would occur once in 14 days, in the proposed circuit before it has been constructed, and the simulator can also be used to find such weaknesses in an existing circuit. One well-known simulator is PULSAR, from Number One Systems Ltd (Oak Lane, Bredon, Tewkesbury GL20 7LR, UK). This has the advantage that it can be integrated with circuit diagram and PCB layout software.

Signature analyser

A signature in this sense means a hexadecimal word, and the signature analyser is a method of finding an error in any logic circuit, including the memory chips for read-only memory (ROM), erasable programmable read-only memory (EPROM) or CMOS random access memory (RAM). The action is that a hexadecimal word is produced from each output when a set of inputs is applied to all the possible inputs of the circuit. This is a more specialized instrument than the others in this section.

The analyser uses a serial register with feedback taken from four of its parallel outputs to an XOR input. The feedback is applied so that any typical set of inputs will produce a set of bits in the register which is almost certainly unique. The principle, based on information theory, is that a register is working most efficiently when all states are equally likely (an n-bit shift register produces $2n - 1$ different output patterns).

Without feedback, a 16 bit register would produce a hex word that would simply reflect the last 16 bits fed into it. For example, if you fed in 16 1s, the register word would be FFFFH, and this would not alter if you continued to feed in 1s. By contrast, with the feedback connections illustrated, the action of feeding in 1s will produce a different register word for each 1 bit fed in, and there will be no repetition until 65536 bits have been fed in. This register arrangement is also used to produce 'random' numbers, and when larger registers are used the chance of recurring numbers becomes less. In the 16 bit example, the words that are generated from 65, 536 or fewer entries are therefore unique, and this allows us to trigger the start and stop of a signature analyser from the highest order of address line (A15) of an 8 bit microprocessor system, or from the A15 line of a system that uses more than 16 address lines.

The data input to the analyser will be from any line that produces signals, so that address or data lines can be used. Contact can be made by a probe or a clip. The register is also gated so that it operates for a fixed window, a defined number of clock cycles. In use, the system under test is 'exercised', meaning that inputs are applied, preferably so that all possible signal values are used at the inputs. The register of the signature analyser will fill with bits in the gated time window, and when the gates have closed, the four hex numbers displayed by the register form a word that should be unique to that combination of inputs and the point that is being tested.

The signature analyser must be used along with software that ensures that the inputs will be consistent, and a system that is known to be perfect is used to provide the signature words. If a system on test displays a different signature this will point to a fault, and the position of the fault can be found by repeating a signature check at different points along a signal path. One merit of signature analysis is that the same type of fault at a different point in a circuit will usually produce a different signature word.

Testing of combinational circuits is straightforward, but sequential circuits can be tested only if all feedback connections are broken.

Microprocessor circuits are usually signature tested by inserting an adapter between the microprocessor and its socket. The adapter is designed to allow pins to be open-circuited or connected to either logic level so as to prevent interrupts or to isolate data lines from memory, and also to set a fixed data input such as the no-operation (NOP) code which will allow the clock to cycle the address lines continually. The probe can then be used at each point where a signature is known.

Figure 39.2 illustrates the modifications that have to be made to a Z80 microprocessor circuit to carry out signature analysis. The interrupt line is broken so that the interrupt pin can be taken high, and the data bus is also broken so that the NOP command is always on the microprocessor data pins. These changes can be incorporated into an adapter that fits between the microprocessor chip and its normal socket.

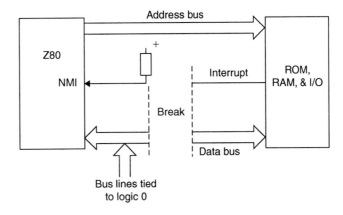

Figure 39.2   Signature analysis modifications to a Z80 circuit

A more restricted meaning of signature is applied to the ROM of a computer. The signature can be found in a variety of ways, such as by adding up all the stored byte numbers and taking the remainder after dividing by some factor; whatever method is used will have been devised so that the resulting word is unique and will be found only if the ROM contents are uncorrupted. The word that is obtained in this way will also be stored in the ROM, so that if the two do not match there must be a ROM error. This system can be used also on RAM when the RAM is filled with specified bytes.

Practical 39.3

Use suitable test equipment to locate faults in systems that make use of combinational logic circuits.

Multiple-choice revision questions

39.1 Why does a service engineer need a good grasp of principles of electricity and electronics?
(a) it shows a good general education
(b) it makes it easier to deal with customers
(c) it makes it easier to use the Internet
(d) not all faults are documented; you need to figure them out for yourself.

39.2 Half-split or divide-and-conquer is a method of
(a) testing bits of a circuit at random
(b) logically finding where a fault is located
(c) dividing servicing responsibility between engineers
(d) working with circuits on more than one board.

39.3 Negative feedback loops present a servicing problem because
(a) the equipment was originally set up without a feedback loop
(b) the loop might cause oscillation
(c) the loop cannot be disconnected
(d) you cannot be sure where a fault within the loop might lie.

39.4 Digital systems require specialized equipment because:
(a) signals have large bandwidths
(b) pulse signals may occur only at intervals
(c) rise and fall times cannot be measured with an analogue oscilloscope
(d) pulse shapes cannot be seen on an analogue oscilloscope.

39.5 The most common problem with circuit boards is:
(a) excessive flux on the board
(b) excessive solder on the board
(c) use of surface-mounted components
(d) cracks and dry joints.

39.6 One main advantage of using a digital multimeter is:
(a) quick reading
(b) high input resistance
(c) it can follow rapid voltage changes
(d) more accurate reading.

Answers to multiple-choice questions

Chapter 26   26.1(c); 26.2(c); 26.3(b); 26.4(c); 26.5(b)

Chapter 27   27.1(c); 27.2(b); 27.3(c); 27.4(b); 27.5(b)

Chapter 28   28.1(b); 28.2(c); 28.3(b); 28.4(a); 28.5(c); 28.6(c)

Chapter 29   29.1(b); 29.2(b); 29.3(c); 29.4(b); 29.5(a)

Chapter 30   30.1(a); 30.2(b); 30.3(d); 30.4(a); 30.5(b); 30.6(c)

Chapter 31   31.1(c); 31.2(b); 31.3(b); 31.4(d); 31.5(c); 31.6(c)

Chapter 32   32.1(c); 32.2(a); 32.3(c); 32.4(b); 32.5(c)

Chapter 33   33.1(c); 33.2(a); 33.3(c); 33.4(b); 33.5(c); 33.6(b)

Chapter 34   34.1(c); 34.2(a); 34.3(c); 34.4(b); 34.5(b); 34.6(d)

Chapter 35   35.1(c); 35.2(b); 35.3(d); 35.4(b); 35.5(c)

Chapter 36   36.1(c); 36.2(b); 36.3(c); 36.4(d); 36.5(b)

Chapter 37   37.1(b); 37.2(d); 37.3(b); 37.4(c); 37.5(a)

Chapter 38   38.1(b); 38.2(c); 38.3(c); 38.4(b); 38.5(a); 38.6(c)

Chapter 39   39.1(d); 39.2(b); 39.3(d); 39.4(b); 39.5(d); 39.6(b)

Index

Printed in the United States
106989LV00002B/60/A